III-Nitride Semiconductors: Lighting the Way to the Future

Navya

TABLE OF CONTENTS

Chapter 1 20

Introduction 20

1.1.   **Background and Overview** 20

1.2.   **Applications** 22

1.3.   **Objective and Content of the** book 24

1.3.1.   Objective 24

1.3.2.   Content 24

Chapter 2 Material System 26

2.1.   **Semiconductor Bandgap Structure** 26

2.2.   **III-Nitride and GaN Crystal Structures** 28

2.3.   **Optical Properties and Recombination Dynamics** 30

2.3.1.   Luminescence Properties of III-Nitrides 30

2.3.2.   Radiative Recombination Process 31

2.3.3.   Non-Radiative Recombination Process 33

2.3.4.   External and Internal Efficiency 35

2.3.5.   Radiative Efficiency 36

2.3.6.   Extrinsic Efficiencies 36

2.3.7.   The relationship between efficiency and a radiative and total lifetime 38

2.4. Confinement Effect in the III-Nitride Nanostructures 38

2.4.1.   Density of State and Confinement Conditions in Nanostructures 38

2.4.2.   Optical Properties of III-Nitride Quantum Well 40

2.4.3.   Optical Properties of Quantum Dots 42

2.5.   III-Nitride Growth Techniques 43

2.6.   The Kinetic Energy and Surface Energy Effects on Nanostructure Formation 43

2.6.1.   The Surface Energy 43

2.6.2.   Growth Kinetics Energy 45

2.7.   Threading Dislocation (TDs) Defects in III-Nitrides 46

2.8.   Challenges Facing III-nitride and GaN Growth and Fabrication 48

2.9.   Functionalizing GaN with Emerging Materials 49

2.9.1.   Perovskites Materials 50

2.9.2.   Wide Band Gap MnO QDs 53

2.10.   Optoelectronic Device Operational Mechanisms 53

2.10.1.   Photodetectors 53

2.10.2.   Light Emitting Diodes (LEDs) 56

Chapter 3 ....**58**

Experimental Techniques **58**

3.1.   Material Growth and Fabrication Techniques 58

3.1.1.   Pulsed Laser Deposition Technique 58

3.1.2.   Femtosecond-Laser Ablation in Liquid 59

3.1.3   Magnetron Sputtering 61

3.2.   Structural Characterization 62

3.2.1.   X-ray Diffraction 62

3.2.2.   Scanning Electron Microscopy 63

3.2.3.   Transmission Electron Microscopy 65

3.3.   Optical Characterizations 67

3.3.1.   Photoluminescence Measurement 67

3.3.2.   Photoluminescence Excitation (PLE) 69

3.3.3.   Time-Resolved Photoluminescence Spectroscopy 70

3.3.4.   UV-VIS Absorption Measurements 72

3.4.   Complementary Techniques Used by Collaborators 73

3.4.1.   Cathodoluminescence 73

3.4.2. Focused Ion Beam (FIB) 74

3.4.3. Molecular Beam Epitaxy (MBE) 75

Chapter 4 77

High-quality self-assembly GaN nanowires (NWs) grown on different substrates by using pulsed laser deposition 77

4.1. Introduction 77

4.2. Experimental method 80

4.2.1. Structural Characterizations 80

4.2.2. Optical properties 80

4.3. Results and discussions 80

4.3.1. Substrates used in this work 80

4.3.2. High-quality GaN NWs growth 81

4.3.3. Optimizing PLD conditions 83

4.3.4. Characterizations 92

4.3.5. Large scale growth (4-inch wafer) 96

4.4. Summary 98

Chapter 5 99

Growth Mechanism of high-optical efficiency dislocation-free GaN NWs without catalyst or seeding 99

5.1. Introduction 99

5.2. Experimental method 100

5.2.1. The sample preparation 100

5.2.3. Structural Characterizations 100

5.2.3. Optical characterizations 101

5.3. Results and discussions 101

5.3.1. Advanced High-resolution STEM Analyses 101

5.3.2. Growth mechanism of single-crystal NWs via *in-situ* polycrystal-line WL using the surface energy concept 110

5.3.3. High optical efficiency 113

5.4. Summary 119

Chapter 6 121

Light-emitting devices structure based on PLD GaN NWs 121

6.1. Introduction 121

6.2. LED structure based on PLD GaN NWs on p-GaN 123

6.2.1. LED Device Fabrication 123

6.2.2. I–V Characterization 124

6.3. LED structure based on InGaN/GaN MQW layers grown on PLD GaN NWs 125

6.3.1. Characterization 127

6.4. Summary 131

Chapter 7 132

Comparative investigation: Enhanced Photodetectors based on GaN nanowires functionalized by different perovskites via work function engineering 132

7.1. Introduction 132

7.2. Experimental Method 134

7.3. Results and Discussion 136

7.3.1. Materials Properties 136

7.3.2. Device characteristics 139

7.4. Summary 151

Chapter 8 153

Enhanced UV efficiency of GaN nanowires functionalized by wider bandgap solution-processed p-MnO quantum dots via the Energy transfer process 153

8.1. Introduction 153

8.2. Experimental Method 156

8.2.1. p-type MnO QDs Synthesis 156

8.2.2. GaN NWs Synthesis 156

8.2.3. Structural Characterizations 156

8.2.4. Optical Characterization 157

8.3. Results and Discussion 158

8.4. Summary 171

Chapter 9 173

Conclusion and Future Direction 173

9.1. book Conclusion 173

9.2. Future Direction 175

9.2.1.   Material Growth 175

9.2.2.   Device Fabrication 176

Chapter 1

Introduction

1.1. Background and Overview

III-nitride semiconductors (including their alloys) are considered the most important materials for optoelectronic applications due to their unique properties: (i) III-nitride bandgap is tunable and direct and the bandgap energy (E_g) values vary from $E_g = 0.7$ eV for InN, to $E_g = 6.2$ eV for AlN, while $E_g = 3.4$ eV for GaN, which is ideal for efficient emitters.[1] (ii) III-nitrides are characterized by high chemical and thermal stability due to their strong chemical bonding, resulting in high melting points and high mechanical strength.[2] Thus, they are highly resistant to radiation damage or electrical degradation related to high currents that can occur in the III-nitride active regions of electrical or emitting devices. (iii) Unlike II−VI semiconductors, these materials are also characterized by good thermal conductivity,[3] resulting in devices that can operate in harsh environments, and at higher temperatures.[4] (iv) High electron mobility, high saturation velocity, and high breakdown field, are considered as the key advantages associated with III-nitrides, particularly, if these materials are intended for high-speed and high-power electronic applications.[5]

GaN growth was initiated in the first decades of the 20th century,[6] preceding Si semiconductors. However, the difficulty of III-nitride synthesis prevented any progress in the use of this material until the 1960s. This changed when, in 1969, J. Tietjen and his colleagues used the hybrid vapor phase epitaxy (HVPE) technique for the first time to successfully grow a GaN film.[7] This work was followed by the first GaN growth using metalorganic chemical vapor deposition (MOCVD) by H. Manasevit, and molecular beam epitaxy (MBE) growth by I. Akasaki, in 1971[8] and 1974,[9] respectively. Nevertheless, at that time, bulk crystal technology was not sufficiently evolved. Indeed, despite considerable scientific interest in LEDs which started before 1907, it was not until 1971 that Pankove and his co-workers developed the first III-nitride LED.[10] Following the introduction of the low-temperature nucleation layer concept in 1983,[11] a remarkable breakthrough in III-nitride technology was achieved by the demonstration of p-type GaN in 1989.[12] Two research groups led by Amano and Akasaki have simultaneously conducted extensive work on GaN-based light emitters, whereas the commercialization of the blue and green LEDs was achieved by Nakamura and his colleagues.[13] The most significant developments in the field of III-nitride-based semiconductor technology took place in the 1990s. Once the researchers were capable of producing high-quality III-nitrides, understanding, as well as controlling, their optical properties became the driving force for the realization of optoelectronic devices. However, as obtaining commercial native substrates of sufficient quality, and required quantity for nitride growth is prohibitively costly and very difficult, researchers have thus far depended on hetero-epitaxy for device fabrication—a crystal growth using a foreign substrate. Owing to the remarkable breakthrough that has been achieved in this field, devices based on heteroepitaxial III-nitrides fabricated on SiC substrates and sapphire are now commercially available.[14] Extensive research efforts have also been dedicated to other substrates, such as GaAs, Si, $LiGaO_2$ and AlN, with the goal of obtaining high-quality GaN materials.[15]

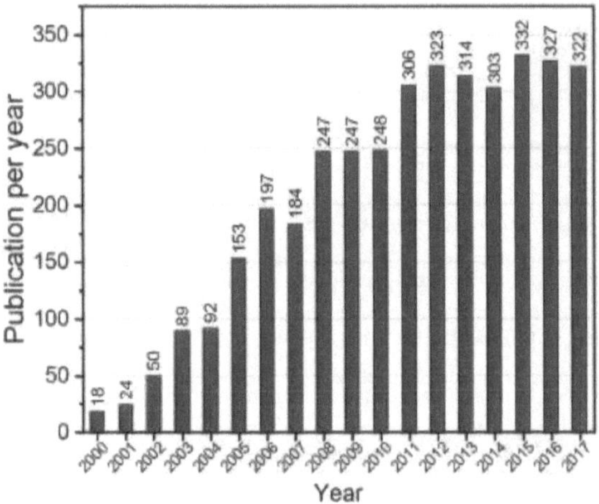

Figure 1.1. The number of publications pertaining to GaN NWs since 2000, based on Web of Science search results. *Source Science citation index (Web of Science), ISI Web of Knowledge: search date January 15th, 2020.*

1.2. Applications

A significant number of current and potential optoelectronic and electronic high power, high frequency, and photovoltaic applications (are?) based on GaN and other III-nitrides. *Figure 1.2 shows the projections indicating that the global gallium nitride (GaN) semiconductor device market is expected to reach USD 4.37 billion by 2025, while,* for example, *the global GaN power device market is likely to achieve a CAGR of 79% during the 2017–2022 period. It is expected that this market value will reach US$ 460 million at the end of the period.*

The energy range associated with wide bandgap is beneficial for photovoltaic devices, as they can work as absorber layers, given that their absorption edge can be easily differentiated for optimizing cell efficiency.[16] Thus, III-nitrides (including GaN) are the best candidates for optoelectronic and electrical applications,[15, 17] such as light-emitting diodes (LEDs), laser diodes (LDs), and electric power devices operating in high-power and high-frequency domains for sensing, medical curing, and industrial applications in harsh environments as shown in Figure 1.3.[18-22]

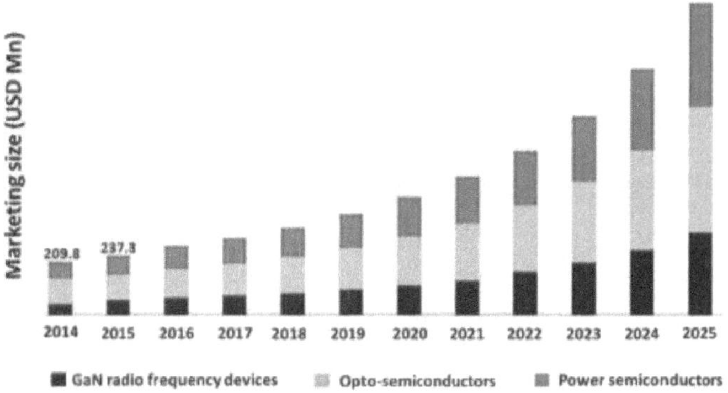

Figure 1.2. The projections indicating the global gallium nitride (GaN) semiconductor device market from 2014 to 2025.[23]

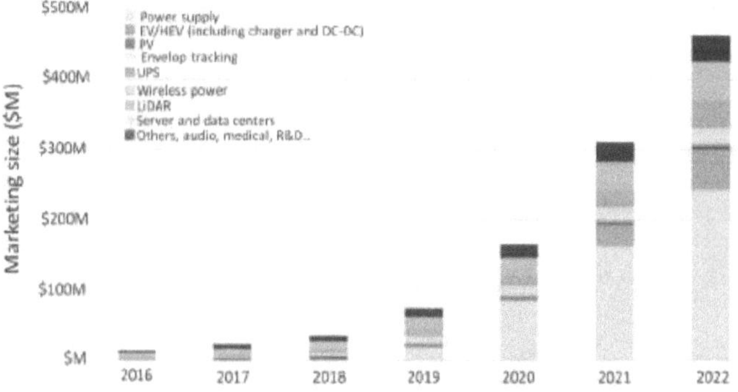

Figure 1.3. Histogram for marketing size of GaN applications per year.[24]

In particular, III-nitride nanowires (NWs) have attracted considerable attention from researchers in recent years (as shown in Figure 1.1) owing to their superior physical properties compared to their bulk counterparts (microstructure materials).[19] These features have motivated research into their application in the fabrication of optoelectronic and electronic devices. In particular, NWs composed of group III−nitride elements have huge potential for use in revolutionary semiconductor device configurations.[19] Specifically, GaN NW is one of the best candidates for the development of optoelectronic devices and is already used in the manufacture of advanced nanoscale devices.[25] Thus far, GaN NWs have been used in the form of key elements in multi-color and single nanoscale LEDs; in optically, as well as electrically-driven lasers; and in polarization-sensitive photodetectors.[25]

1.3. Objective and Content of the book

1.3.1. Objective

The work reported in this book aims to develop a cost-effective growth strategy for fabricating dislocation-free GaN nanowires (NWs) on a wide range of substrates. This includes conductive and/or transparent bulk materials as emerging 2D substrates by pulsed laser deposition (PLD) for optoelectronic devices, as well as hybridizing the GaN NWs with emerging materials to enhance the optical efficiency and advance the performance of GaN-based devices.

1.3.2. Content

This book contains eight chapters. A brief background and objectives are presented in Chapter 1. Chapter 2 is designated for a brief introduction to III-nitrides (focusing on

GaN mainly) properties and their applications, and nanostructure growth mechanisms. A summary of emerging material properties (quantum dots and perovskite) that can be used in GaN-based devices is also presented in this chapter. Experimental techniques used in the projects reported in this book are briefly described in Chapter 3. The results obtained as a part of the present study are reported in Chapters 4, 5, 6, 7, and 8 along with the discussions of the key findings. Focus is given to exploring the effect of the growth parameters to develop a novel GaN NW growth strategy on any bulk and 2D substrates without catalyst or seeding using a one-step PLD method. Chapter 5 is reserved for presenting the high optical and structural properties of optimized vertically aligned dislocation-free GaN NWs and discussing the related growth mechanisms using advanced optical, structural, and electrical characterizations. Light-emitting device structures based on these GaN NWs, including LED-based on MQWs grown on NWs, are presented in Chapter 6. An investigation into photodetector properties that are based on GaN NWs functionalized by different perovskite materials is reported in Chapter 7. A new strategy to enhanced GaN NW optical efficiency by functionalizing NWs by metal-oxide quantum dots is presented in Chapter 8. Chapter 9 concludes the book with a key study finding, followed by planned future work, to advance the presented projects.

Chapter 2

Material System

In this chapter, a brief description of the semiconductor bandgap structure is presented, followed by an outline of III-nitride (mainly GaN) materials and their properties, including nanostructure properties. The crystal structure and the optical properties of GaN are also introduced, and the emerging materials, such as perovskite oxide quantum dot, used in the reported studies for functionalizing GaN devices, are briefly outlined. Finally, the chapter closes with a brief description of physical growth mechanisms and device working principles.

2.1. Semiconductor Bandgap Structure

Bandgap, also known as the energy gap (E_g), is conventionally defined as the difference in the energy of the conduction band (CB) minimum and that of the valence band (VB) maximum.[26] The semiconductor bandgap structure is a key parameter as it controls the optical and electrical properties, while the nature of crystal bonding affects both the bandgap structures and the other crystal structure and material properties. Bandgaps are inversely proportional to the interatomic spacing (i.e., bond length). [27] Bandgap energies are the most important characteristics of semiconductors, as their respective band structures determine the range of their potential applications.

Based on their band structure, semiconductors can be categorized into direct and indirect bandgap materials. For direct bandgap materials, the VB maximum and the CB minimum occupy the same crystal momentum (mainly at k = 0), as shown in Figure 2.1a, which is not the case for indirect bandgap semiconductors, as shown in Figure 2.1b.[28]

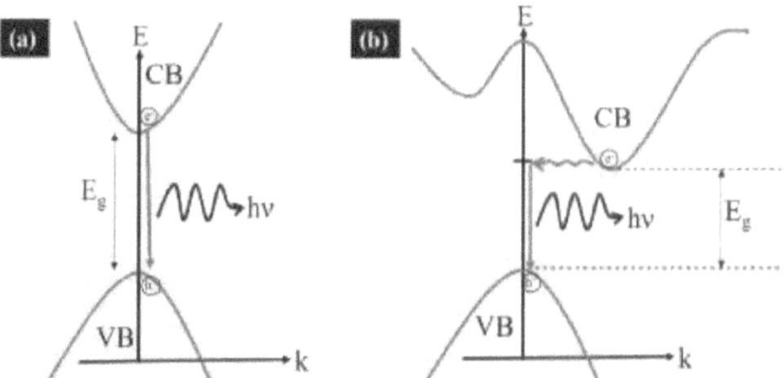

Figure 2.1. Band structures of (a) direct bandgap semiconductor and (b) indirect bandgap semiconductor. The carrier excitation and relaxation pathways for both structures (the red arrow is the phonon energy to conserve the momentum) are shown.

III-nitrides, including GaN, are direct bandgap semiconductors,[29] as shown in Figure 2.2. Figure 2.3 depicts the inverse relationship between the interatomic spacing (lattice constants) and the bandgap of III-nitride materials, indicating that this material and its alloys operate from the deep UV to the infrared (IR) region of the electromagnetic spectrum due to bandgap tunability.

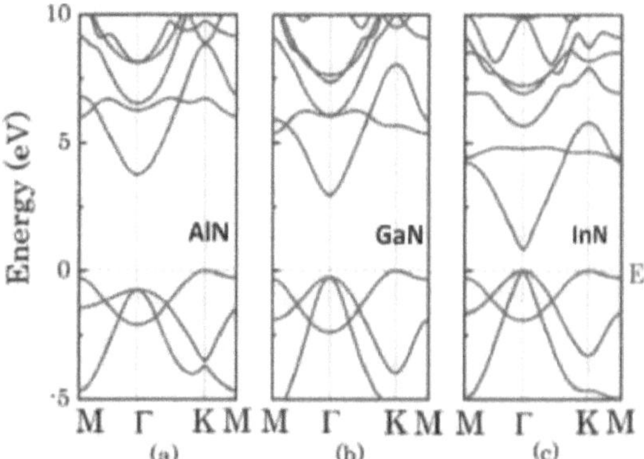

Figure 2.2. Band structures of different III-nitride materials (direct bandgap materials).[30]

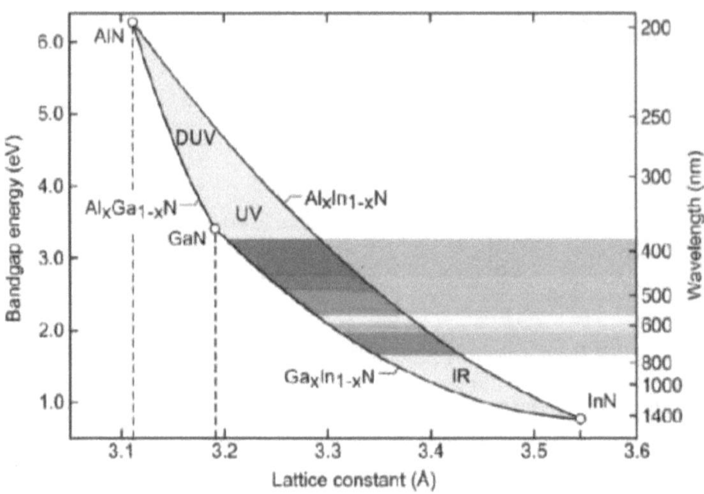

Figure 2.3. Bandgap energy as a function of lattice constants for III-nitride materials and their alloys, resulting in the tunable bandgap.[31]mma

2.2. III-Nitride and GaN Crystal Structures

III-nitride semiconductors are thermodynamically most stable in the wurtzite (hexagonal) structure. However, under certain growth conditions, this material can be crystallized in zinc-blende (cubic) or rocksalt (unstable) structure.[32] Figure 2.4 shows the wurtzite (wz), where c and a represent lattice parameters and zinc-blende (zb) structures of GaN. The hexagonal and cubic phases vary only in terms of the accumulating sequence concerning close-packed planes.[33] The structure of zinc-blende belongs to the *F43m* space group, which includes an ABCABC stacking sequence, while the structure of wurtzite has an ABABAB stacking sequence, which belongs to the space group named *P63mc*, as shown in Figure 2.4.[34] The energy bandgap and the lattice parameters for these two III-nitride structures are shown in Table 2.1.

Wurtzite structure lacks inversion symmetry, i.e., the centers of positive, and negative charges do not correspond. The absence of a center of symmetry in wurtzite structure, and the fact that the III−N bond has a strong ionic character, result in unprompted polarization. Also, the lack of inversion symmetry gives rise to special properties resulting in polarization, such as piezoelectricity (the phenomenon whereby externally applied strain in a crystal structure results in the production of piezoelectric polarization).[35] Even though several properties of III-nitrides are well documented, it is worth noting that, through alloying GaN (3.4 eV) with InN (0.7 eV) and/or AlN (6.2 eV), the bandgap can be aligned in a controllable manner, which can lead to novel optoelectronic and electronic applications.

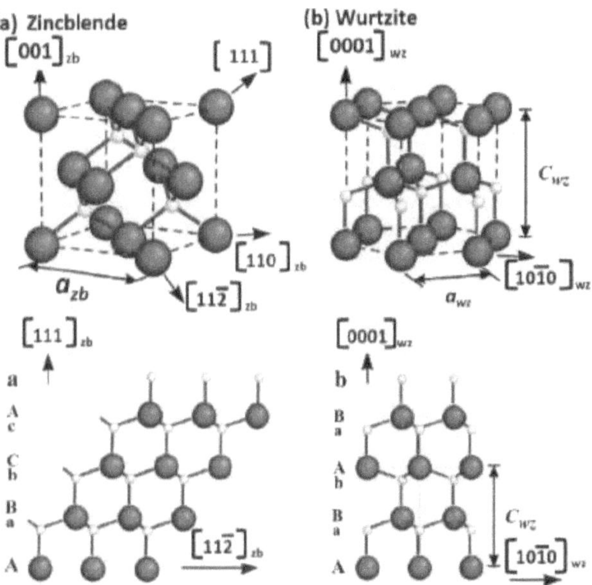

Figure 2.4. Wurtzite (wz) and zinc-blende (zb) structure configurations of GaN.[34]

Table 2.1. Energy bandgap values and lattice parameters for wurtzite and zinc-blende structures of different III-nitrides at room temperature (RT).[33, 36]

Property	AlN		GaN		InN	
	W	zb	W	zb	W	zb
Energy Gap (eV)	6.5	5.4	3.39	3.299	0.7	0.78
Lattice constants (Å)	[a]3.111 [c]4.979	4.38	[a]3.189 [c]5.185	4.50	[a]3.538 [c]5.703	4.98

2.3. Optical Properties and Recombination Dynamics

2.3.1. Luminescence Properties of III-Nitrides

When samples are subjected to photon, electron, or heat energy that exceeds that of the material bandgap, electron carriers in the VB are excited to the CB levels, leaving hole carriers in the VB. These exciting carriers (electrons and holes) recombine by either radiative or non-radiative recombination processes. Radiative recombination, whereby photons are emitted after the recombination, is typically observed in the luminescence spectrum produced by the material. On the other hand, non-radiative recombination occurs when carriers are transferred within small energy levels (corresponding to the infrared spectral region). Thus, in this process, heat is generated as a result of carrier-phonon (lattice vibration) interactions, which is undesirable for optoelectronic devices.[37]

2.3.2. Radiative Recombination Process

In the radiative recombination case, there are three radiative paths—(i) band-to-band (band edge) emission, (ii) donor−acceptor pair (DAP) recombination, and (iii) defect-related emission.

i)   *Band-to-band (band edge) emission*

Band-to-band (band edge) emission occurs when the excited electron returns directly to the VB by releasing energy that is roughly equal to the direct bandgap. When an excited electron in the CB is attracted to a hole in the VB by the electrostatic Coulomb force, this results in a bounded electron−hole pair known as "exciton," as shown in Figures 2.5a and 2.5b. If the excitons can move freely within the crystal with a large exciton radius, they are referred to as *free excitons* (see photoluminescence (PL) of the (FE, X) peak in Figure 2.5b), whereas *bound excitons* for example donor-bound excitons (D^{o}, X), peak in Figure 2.5b) are localized at the impurity sites with a much smaller radius.[38-40] The excitonic recombination will be dominant at certain low temperatures.[37] At room temperature (RT), the near band edge (NBE) emission of bulk GaN is ~ 3.40 eV, increasing to ~ 3.467 at low temperatures.[41, 42] This emission exhibits a redshift as the temperature increases due to the bond expansion at high temperatures, causing bandgap shrinkage.[38]

ii)   *Donor and acceptor-related emission*

The donor and acceptor pair (DAP) emission arises if the electron−hole recombination from the donor or the acceptor energy level occurs, whereby a photon is emitted with an energy that is lower than the bandgap energy, as shown in Figure 2.5a using the GaN spectrum as an example. Semiconductors doped (intentionally or unintentionally) by acceptor and donor impurities produce DAP emission, mostly at low temperatures, since it is quenched at room temperature. In general, free excitons, bound excitons, and the DAP are considered NBE emission.[38]

iii) *Defect-related emissions*

This emission type involves a broad defect band, such as the yellow luminescence (YL) band that is observed in the GaN emission spectra in the 2.1–2.3 eV range, as shown in Figure 2.5a.[43] This band is therefore used in the evaluation of material quality. The ratio of YL intensity and the NBE integrated intensities is a good indicator of the optical quality of a particular semiconductor.

When an electron recombines with a hole in the radiative recombination process, the recombination rate must be proportional to the number of excess electrons generated per second, i.e., $R_{rad} \propto \Delta n$. Moreover, the number of radiative recombination events should be proportional to the number of excess holes generated per second ($R_{rad} \propto \Delta p$), whereby $R_{rad} \propto \Delta n \, \Delta p$. Since $\Delta n = \Delta p$, the radiative rate equation is given by:[44]

$$R_{rad} = B\Delta n^2; \qquad\qquad 2.1$$

where B is the radiative coefficient of the material.

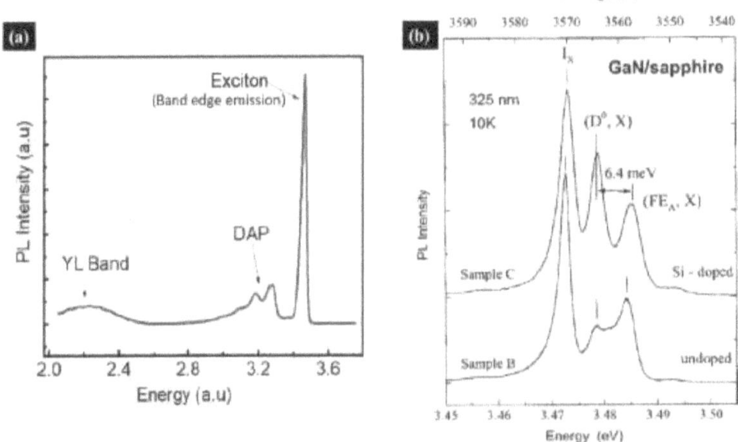

Figure 2.5. PL emission spectra at 10 K of (a) doped GaN and (b) undoped GaN to present donor-acceptor pair (DAP), free exciton (FE, X) and donor-bound exciton (D^0, X) emissions.[39, 40, 43]

2.3.3. Non-Radiative Recombination Process

In the non-radiative recombination process, carrier-phonon interactions and carrier-carrier interactions occur. Their carrier dynamics involve a series of sub-mechanisms, such as (i) carriers being captured by defect sites, known as Shockley-Read-Hall (SRH) recombination, at low excitation power; (ii) Auger recombination; and (iii) carriers escaping thermally through phonon interactions.[45] Figure 2.6 shows the possible recombination transitions and the energy levels for GaN materials.[43]

i) ***Shockley-Read-Hall (SRH) (Trap-Assisted) recombination***

Sometimes, carriers move to the defect levels inside the bandgap, known as deep-level trap (DLT) states (generated by crystal defects), and the excess energy is released through heat or vibrations (phonons). Consequently, the carriers recombine without releasing photons, or their recombination results in photons with energy lower than the bandgap energy[46] as depicted in Figure 2.6c. As only one carrier at a time can occupy DLT, the SRH recombination rate (R_{SRH}) and the density of excess carriers (Δn) are directly proportional, as expressed by the following equation:[46]

$$R_{SRH} = A\Delta n; \qquad\qquad 2.2$$

$$A = \sigma v_{th} N_t; \qquad\qquad 2.3$$

where σ represents the capture cross-section of the DLT, v_{th} indicates the carriers' thermal velocity, and N_t is the DLTs' density in the material.

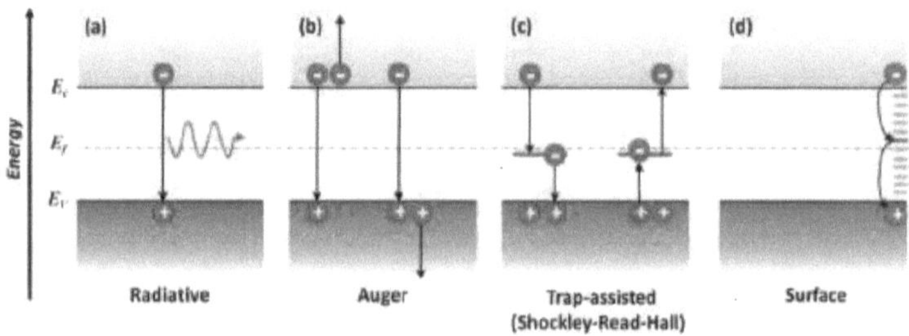

Figure 2.6. Possible recombination transitions and the bandgap energy levels in III-nitride materials.[47]

ii) *Auger recombination*

Auger recombination was considered as the main cause of the optical efficiency droop in emitting semiconductor devices.[48, 49] During this process, an electron and a hole recombine and pass on the excess energy to another carrier instead of emitting a photon,[50] as shown in Figure 2.6b. Figure 2.7a and 2.7b depict a type of Auger recombination rate (R_{Auger}) that consists of a three-carrier process involving either electron–electron–hole (eeh) or electron–hole–hole (ehh) interactions (direct Auger), which can be expressed as:[50]

$$R_{Auger} = c\Delta p \Delta n^2 = c\Delta n \Delta p^2 = \Delta n^3;\qquad 2.4$$

where c is a constant. Indirect Auger recombination can occur through phonon emission, resulting in total momentum conservation, as shown in Figure 2.7b.[50]

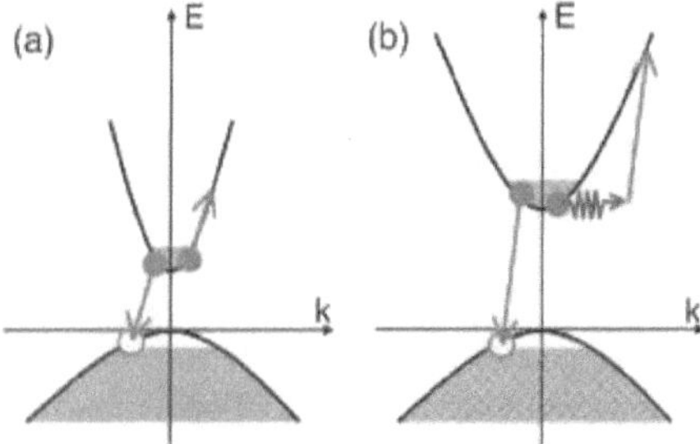

Figure 2.7. Schematic diagram of direct and indirect Auger recombination.[51]

2.3.4. External and Internal Efficiency

External quantum efficiency (EQE) represents the ratio of the number of carriers collected by the device to the number of photons incident on the device, while internal quantum efficiency (IQE) is defined as the ratio of the number of carriers collected by the device to the number of photons absorbed by the device.[38, 52] These efficiencies are linked via the following expression:[53]

$$IQE = \frac{EQE}{LEE};\qquad 2.5$$

where LEE is light extraction efficiency and LEE= 1−Reflectance. Typically, IQE is higher than EQE because reflectance is always positive, thus ensuring that (1-reflectance) < 1 holds for any device.

2.3.5. Radiative Efficiency

Combining Equation 2.1, 2.2, and 2.4 respectively, by the three recombination processes described earlier, yields the well-known "ABC recombination model", which can be related to the generation carrier rate, G_T, as follows:[54]

$$R_T = R_{SRH} + R_{rad} + R_{Auger} = A\Delta n + B\Delta n^2 + C\Delta n^3 = G_T; \qquad 2.6$$

where R_T is the total density of carriers recombined in unit time.

The efficiency of any system is defined as the ratio of its useful output to input. This same principle is applied to the intrinsic radiative efficiency of the material, known as the internal quantum efficiency (IQE or η_{int}), which is given by [55]

$$IQE = \frac{B\Delta n^2}{G_T} = \frac{B\Delta n^2}{A\Delta n + B\Delta n^2 + C\Delta n^3}; \qquad 2.7$$

2.3.6. Extrinsic Efficiencies

In general, all photons emitted by the emitting devices cannot be completely collected due to many reasons, such as reabsorption of a fraction of emitted light by traps inside the semiconductor, absorption by metal contact layers, and total internal reflection. Thus, light extraction efficiency (LEE) is introduced to quantify the ability of a device to supply useful light. It is defined as follows:[55]

$$LEE = \frac{\text{number of collected photons per second}}{\text{photons generation rate}} = \frac{P_{out}/h\nu}{Bn^2} \qquad 2.8$$

Owing to the above relationship in Equation 2.5, EQE is obtained by multiplying IQE and LEE, thus:[55, 56]

$$EQE = IQE \times LEE; \qquad 2.9$$

Equation 2.9 and 2.6 show that using the ABC model given above, the IQE can be explained in terms of the carriers generated by an optical pump (G_{opt}) as follows:[57]

$$G_{opt} = P_{laser}(1 - R)\alpha/A_{spot}h\nu; \qquad 2.10$$

where P_{laser} represents the laser power incident on the GaN sample, R denotes the Fresnel reflection at the sample surface, A_{spot} is the area on the sample surface exposed to the laser beam, $h\nu$ represents the incident energy, and α is the GaN absorption coefficient at the emission wavelength.[58] As integrated PL intensity can be defined as $I_{PL} = \beta B n^2$ (where β represents a constant determined by the total collection efficiency of the PL and the exciting active region volume), Equation 2.6 can be rewritten in terms of I_{PL}, yielding:

$$G_{opt} = A\sqrt{\frac{I_{PL}}{\beta B}} + \frac{I_{PL}}{\beta} + C\left(\frac{I_{PL}}{\beta B}\right)^{3/2}; \qquad 2.11$$

$$G_{opt} = Q_1\sqrt{I_{PL}} + Q_2 I_{PL} + Q_3(I_{PL})^{3/2}; \qquad 2.12$$

where Q_1, Q_2, and Q_3 indicate the fitting parameters explained in terms of A, B, C, and β. When plotting I_{PL} versus G_{opt}, the fitting coefficients can be determined by applying Equation 2.12 to the experimental data. Consequently, the IQE at steady state can be expressed as follows:

$$\eta_{IQE} = \frac{Bn^2}{G_{opt}} = \frac{Q_2 I_{PL}}{G_{opt}} \; ; \qquad 2.13$$

2.3.7. The relationship between efficiency and a radiative and total lifetime

The IQE can be calculated in terms of a total lifetime (τ_{total}) and radiative lifetime (τ_{rad}) as follows:[59, 60]

$$\eta_{IQE} = \frac{1/\tau_{rad}}{1/\tau_{total}} \; ; \qquad 2.14$$

Total lifetime can be expressed in terms of a radiative lifetime (τ_{rad}) and non-radiative lifetime (τ_{non}), as shown in the following equation: [59, 60]

$$\frac{1}{\tau_{total}} = \left(1/\tau_{rad}\right) + \left(1/\tau_{non}\right) ; \qquad 2.15$$

Finally, the recombination rate (R) can be obtained, as it is inversely proportional to the lifetime:

$$R = \frac{1}{\tau} \; ; \qquad 2.16$$

2.4. Confinement Effect in the III-Nitride Nanostructures

2.4.1. Density of State and Confinement Conditions in Nanostructures

Exciting semiconductors allows electron carriers in the CB and hole carriers in the VB to move freely within the crystal structure whereby, depending on the material dimensions, carrier movement can be restricted to one, two, or three dimensions (1D, 2D, 3D) by spatial confinement. As the dimensions are in the order of magnitude of the Bohr exciton radius

(< 10 nm), these geometric confinement restrictions introduce a quantum-size effect, which quantizes the continuous bulk energy bands (both CB and VB) into discrete levels, as the density of states becomes discrete, as shown in Figure 2.8.[61] The density of states, as well as the number of the available states, in any material system change according to the spatial dimensions of the confinement.[62] The 2D confinement structure is known as a quantum well (QW), whereas the 1D and 0D confinement structures are referred to as quantum wires and quantum dots, respectively, as shown in Figure 2.8. Table 2.2 shows the number of degrees of freedom corresponding to the dimensionality of the quantum confinement. The density of states as a function of the number of dimensions (3D, 2D, 1D, and 0D) is depicted in Figure 2.8, while Figure 2.9 provides an example of discrete levels for the QW case.

Quantum confinement is important for LED and laser applications, as the excitonic effects are highly prominent in confined structures even at RT, allowing high radiative electron−hole recombination rates while reducing the external disturbance effects of phonon interactions on carrier recombination. At the same time, due to the confinement effects, the electrons and holes remain in closer proximity than in the bulk material, which increases the radiative recombination probability.[63]

Owing to the confined structures, the light emitted by the material can be tuned by controlling the nanostructure geometry. For example, as nanostructure size is reduced, the NBE emission undergoes a blueshift as the separation of the quantized levels increases.[64]

Table 2. 2. The number of dimensions of quantum confinement structures.

Structure	Quantum confinement	Number of free dimensions
Bulk	none	3
Quantum well	1D	2
Quantum wire	2D	1
Quantum dot/box	3D	0

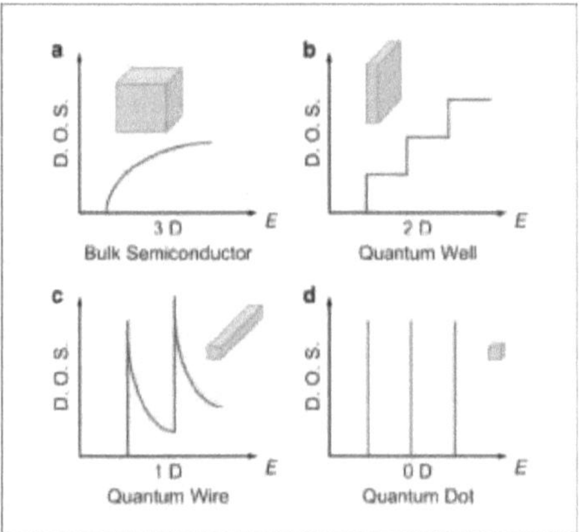

Figure 2.8. The density of states as a function of the spatial dimensions that represent 3D, 2D, 1D, and 0D structures.[62]

2.4.2. Optical Properties of III-Nitride Quantum Well

Quantum well (QW) structure is a heterostructure that can be grown by inserting a very thin layer (2−4 nm) of a semiconductor (well or quantum well) material with a lower bandgap between two outer layers comprising of a semiconductor (barrier or quantum barrier) material with a higher bandgap, as shown in Figure 2.9. This structure exhibits unique optical properties based on the quantum confinement of carriers.[65] The motion of carriers can be quantized in terms of the growth (z) direction, producing a series of distinct energy levels, denoted by dashed lines in the QW illustration shown in Figure 2.9. The 2D confinement behavior of QWs and the quantization energy shift the effective band edge to higher energy ($n = 1$) than the bandgap edge ($n = 0$), thus resulting in a blueshift of the emitted peak (moving the emission to higher energy). This phenomenon allows greater flexibility in structure band engineering, depending on the good width. Inside the QWs, the quantum confinement of the carriers yields much more stable temperature-dependent PL spectra compared to the bulk. As shown in Figure 2.10a, the PL spectrum of bulk GaN at RT is of very low intensity compared to that obtained at low temperature, whereas for the InGaN/GaN MQW emission, the PL intensity is enhanced remarkably at RT, as shown in Figure 2.10b.[66]

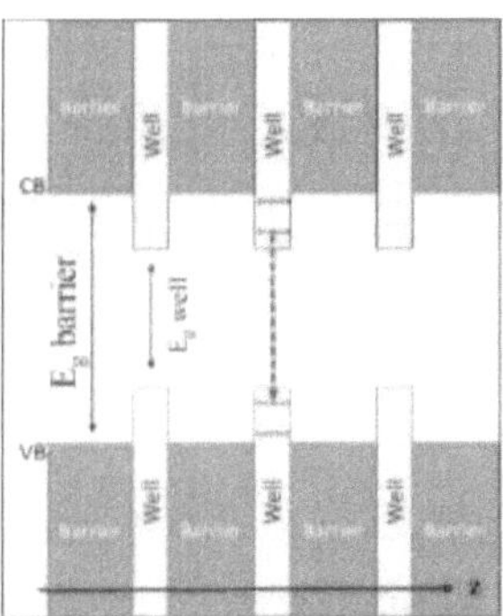

Figure 2.9. Schematic illustration of the MQW structure.

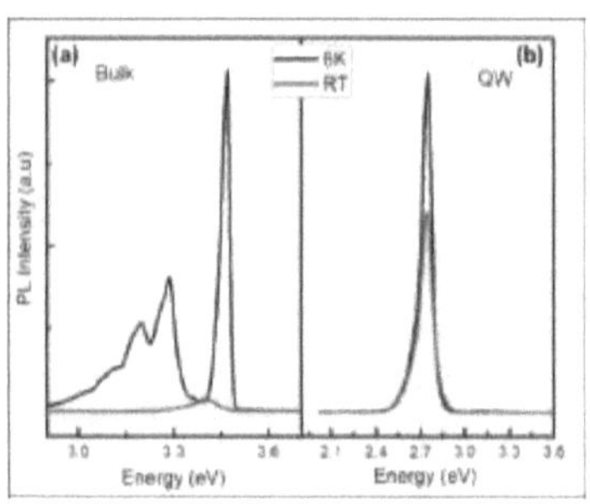

Figure 2.10. A comparison between PL peaks of bulk GaN at (low temperature (black curve and RT (red curve)) in (a) and QWs at (low temperature and RT) in (b).[66]

2.4.3. Optical Properties of Quantum Dots

Quantum dot (QD) is a small fluorescent nanocrystal formed by, for example, a semiconductor, metal, or metal oxide material, usually of spherical shape and <10 nm in diameter.[67] QDs have been at the forefront of research efforts in recent years due to their novel electronic, optical, magnetic, and catalytic properties that differ from those exhibited by larger particles.[67] A QD structure may be considered as a three-dimensional QW with no degrees of freedom.[38] In a QD, the electron and hole states are completely quantized, and the energy spectrum consists of a series of discrete lines, as illustrated schematically in Figure 2.8d. It is to be expected that the increased degree of confinement in quantum dots should yield several benefits. For example, by confining the carriers in all three dimensions, the electron–hole overlap is increased, which in turn improves the radiative quantum efficiency. Furthermore, the discrete nature of the density of states reduces the thermal spread of the carriers within their bands.[38]

2.5. III-Nitride Growth Techniques

Different techniques can be employed for obtaining III-nitride semiconductor structures, such as MQWs and NWs. The most common growth techniques are metal-organic chemical vapor deposition (MOCVD), metal-organic vapor phase epitaxy (MOVPE), and molecular beam epitaxy (MBE), as well as plasma-assisted molecular beam epitaxy (PAMBE).[19, 68]

Pulsed laser deposition (PLD), which is a cost-effective and non-toxic method, has recently been developed as it is a technique with strong potential for use in high-quality semiconductor NW growth,[69] as it overcomes several disadvantages inherent in other techniques, as will be discussed in Chapter 3, Material Growth Section.

In the work reported in this book, high-quality III-nitride materials, in particular GaN NWs that can be grown on any substrate, were obtained for the first time using this technique.

2.6. The Kinetic Energy and Surface Energy Effects on Nanostructure Formation

The two main factors affecting the nanostructure formation are:

- surface energy
- growth kinetics energy

2.6.1. The Surface Energy

Crystal growth using any technique is controlled by the nucleation process and is typically described using different thermodynamic growth modes based on surface energy and supersaturation. Surface energy[70] is dependent on the substrate surface energy (γs), the growing nucleation energy (γn), and the interface energy between the substrate and the growing nucleation (γi). Therefore, three models describing the surface energy in different growth modes have been developed, as shown schematically in Figure 2.11:[71, 72]

I. Film growth mode (*Frank-van der Merwe (FM) mode*): When the sum of the surface energy related to the interface and that which is related to the grown nucleation is equal to the substrate surface energy, i.e., when $\gamma s = \gamma i + \gamma n$ holds, the deposited material completely covers the substrate surface, resulting in layer-by-layer growth, as shown in the schematic diagram presented in Figure 2.11a.[71, 72]

II. The island growth mode (*Volmer-Weber (VW) mode*): When the substrate surface energy satisfies the condition $\gamma s < \gamma i + \gamma n$, deposited material does not fully cover the substrate. In this case, the total energy of the system is decreased by the accumulation of deposited species, resulting in the nucleation of 3D islands, as shown in Figure 2.11b.[71, 72]

III. The island-layer growth mode (*Stranski-Krastanov (SK) mode*): When the substrate surface energy exceeds the sum of the interface surface energy and that of nucleation ($\gamma s > \gamma i + \gamma n$), island formation starts after depositing one or few monolayers. These islands fully cover the substrate surface, forming a granular-like layer or a layer of islands above the substrate, commonly known as the "wetting layer" that assists in 1D island growth, such as NWs, nanorods (NRs), and nanotubes (NTs). This growth mode is facilitated by the stress applied as a result of the lattice mismatch between the substrate and the grown material,[71, 72] as shown in Figure 2.11c.

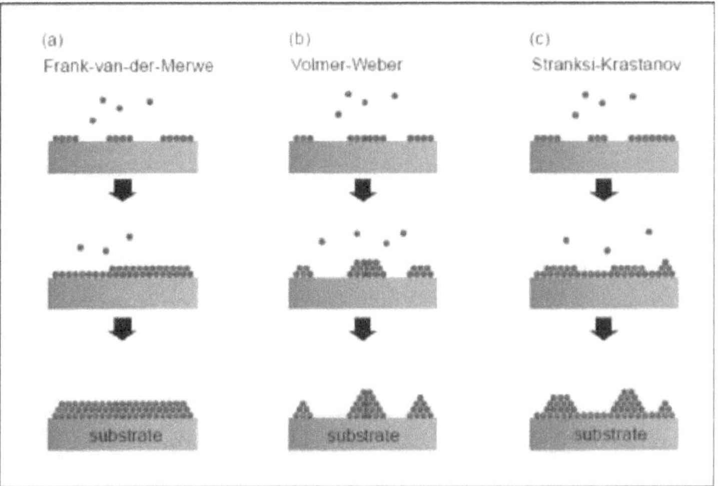

Figure 2.11. Schematic illustration of the three different growth modes (a) Frank-van-der-Merwe (layer-by-layer growth), (b) Volmer-Weber (island growth), and (c) Stranski-Krastanov (island-layer growth).[71]

2.6.2. Growth Kinetics Energy

In PLD, it is possible to change the III-N structure by adjusting the growth kinetic energy of the species (E) as shown below,[72, 73] resulting in NWs:

$$E = E_o \, exp\left(-\frac{d}{\lambda}\right); \qquad 2.17$$

where E_o is the initial charged species energy in the plume emerging from the target surface, d is the distance between the target and the substrate, and λ is the mean free path of the ablated species (the plasma) traveling towards the substrate in the PLD chamber, which depends on the growth temperature (T) and nitrogen pressure (P) and can be calculated using the following expression:

$$\lambda = \frac{kT}{\pi a_o^2 P \sqrt{2}} \; ; \qquad 2.18$$

where a_o denotes gas molecule diameter, and k is Boltzmann constant. Thus, Equation (2.17) becomes:

$$E = E_o \, exp\left(-\frac{\pi a_o^2 P d \sqrt{2}}{kT}\right) \; ; \qquad 2.19$$

This equation explains how the species energy governs the formation of 1D, 2D, or film structures based on the three growth modes mentioned in Section 2.6.1.[72, 73]

In the work reported in this book, based on Equation 2.19, at an optimized pressure, for d = 9 cm, and at an optimized initial species energy E_o (which in this work is equivalent to laser fluence on the sample), the ablated species experience high scattering due to a large number of collisions with nitrogen molecules (represented by a_o in Equation 2.19). As a_o denotes the gas molecule diameter, it is constant for all substrates, as only the N_2 gas environment is used for growth in this work. As only molecules with low E would land on the substrate, island-layer (SK) growth mode (corresponding to the *Stranski-Krastanov* described in Section 2.6.1) would occur. Hence, the nucleation necessary for NW formation can be achieved without catalysis or seeding, irrespective of the substrate type[72, 73], and its lattice mismatch with the grown materials. It is evident from Equation 2.19 that the species energy E does not depend on the substrate type used to form the NW structure, which is confirmed by the findings of the present study, reported in Chapter 4.

2.7. Threading Dislocation (TDs Defects in III-Nitrides

During the growth process, crystal defects are created, manifesting as (i) point defects formed at a single atomic site, such as vacancies, substitutions, and interstitials; (ii) line defects which are formed in a certain direction, such as threading dislocations (TDs); (iii) area defects related to a plane or area (for example, stacking faults, inversion domain boundaries, twins, and grain boundaries); or (iv) volume defects, such as voids, cracks, and atom segregations.[74]

TDs, which are initiated in heteroepitaxial growth on a foreign substrate (shown in Figure 2.12) have an adverse effect on the material quality and device performance, as the lattice mismatch between foreign substrate and grown materials causes the TD formation. Therefore, such defects have been extensively studied. TDs can be defined as sudden changes in the regular ordering of atoms along a certain crystallographic direction. Dislocations may cause highly distorted interatomic bonds, especially in close proximity of the dislocation, which in turn creates minor elastic lattice deformations, resulting in lattice distortion centered around a dislocation line.[75, 76]

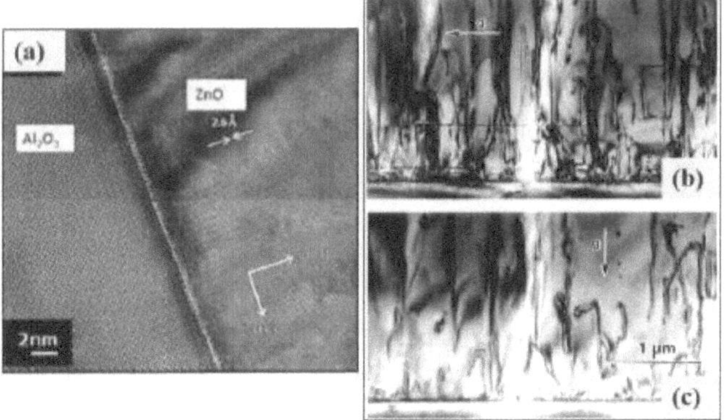

Figure 2. 12. TEM images show (a) dislocation-free interface between the a-sapphire substrate and c-ZnO,[77] and (b and c) high dislocation density along with imaging conditions.[78]

2.8. Challenges Facing III-nitride and GaN Growth and Fabrication

At present, III-nitride growth (GaN mainly) is affected by three main challenges:

(i) Usually, the growth of GaN or III-nitrides is hampered by dislocation defects that are initiated at the material–substrate interface due to lattice mismatch, leading to very low optical efficiency, as shown in Figure 2.13. This is the main obstacle that manufacturers of emitting devices (and III-nitride-based devices in general) still face. Indeed, very famous companies and institutes, as well as the biggest GaN labs in the world led by pioneers in the field, are struggling to improve device efficiency. Consequently, no commercial UV laser diode (operating at $\lambda < 365$ nm) presently exists.

(ii) Although several substrates have very promising characteristics for III-nitride applications and can lead to new functionalities (such as vertical devices, self-powered devices, UV laser devices, etc.), these materials cannot currently be grown on any substrate because of the aforementioned dislocation issue (lattice mismatch) and low material quality, limiting the innovation potential. TDs affect carrier mobility by acting as charge scattering centers.[79] High threading dislocation density (TDD) is thus common in III-nitride semiconductors due to the unavailability of cost-effective foreign lattice-matched substrates (this issue does not affect other semiconductors, such as ZnO[77]), as shown in Figure 2.12a. When there is a substantial lattice mismatch, or/and thermal expansion coefficient difference, between GaN and the commonly used substrates, this causes high TDD,[80] as shown in Figure 2.12b and 2.12c. This would in turn adversely affect the performance of

GaN devices depending on TDs through carrier scattering and non-radiative recombination.[81]

(iii) At present, good-quality GaN and III-nitride growth based on MOCVD or MBE are prohibitively expensive. Moreover, such high-quality materials cannot be grown on different substrates.

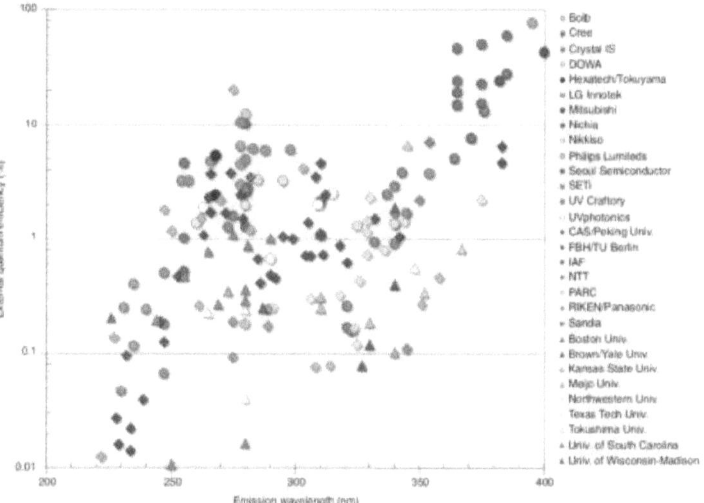

Figure 2.13. EQEs of UV LEDs developed over the last 20 years. Noticeable is the big drop in EQE for wavelengths shorter than 365 nm, which marks the transition from InGaN- to AlGaN-based LED technologies.[22]

2.9. Functionalizing GaN with Emerging Materials

GaN is of particular importance for optoelectronic applications. Therefore, hybridizing (functionalizing) GaN with emerging materials will result in either a new device functionality or would enhance the functionality of existing devices by overcoming some of the issues related to GaN materials. In the work reported in this book, three emerging materials were used to develop GaN-related devices, namely two perovskite materials, along with wide-bandgap QDs. The general properties of these materials are discussed in the subsequent sections.

2.9.1. Perovskites Materials

By hybridizing GaN with perovskite to develop a broadband photodetector, unique advantages can be obtained, such as wide spectral detectivity (spanning the visible and near-IR spectral regions), high charge carrier mobility, and effective light absorption properties.[82] Therefore, combining the perovskite hybrid with the highly stable and UV-sensitive GaN will result in broadband photodetectors suitable for a wide range of applications. In this work, organic/inorganic ($CH_3NH_3PbI_3$) and inorganic halide ($CsPbBr_3$) perovskites were used.

i *Inorganic-Organic Halide $CH_3NH_3PbI_3$ Perovskite Properties and Crystal Structure*

Inorganic-organic halide perovskite Methylammonium lead iodide ($CH_3NH_3PbI_3$), or MAPI, is characterized by a direct bandgap, high mobility, high light absorption coefficient, and high dielectric constant, as well as ability to detect a broad spectrum of visible light. In addition, easy preparation using cost-effective solution-processed synthesis eliminates the need for high temperature or vacuum preparation.[83-85]

The MAPI perovskite is represented by the ABX_3 chemical formula, where A indicates the organic cation (CH_3NH_3), B represents the inorganic cation (Pb^+), and X denotes the halide anion part, such as Cl^-, Br^- and I^-. The bandgap structure can be tuned by changing the halide component (X), thus obtaining specific physical properties required for different bespoke devices or applications.[86] Figure 2.14 shows the typical 3D structure of a MAPI perovskite crystal, which consists of eight octahedral PbI_6 structures with Pb at the center.[87] In the expanded form of the crystal, corners of the PbI_6 octahedral are shared, with the CH_3NH_3 cation at the center enclosed by twelve iodide ions. The MAPI lattice parameters ($a = b = 8.81$ A° and $c = 12.67$ A°) and its space group (14/mcm) have been previously determined.[88]

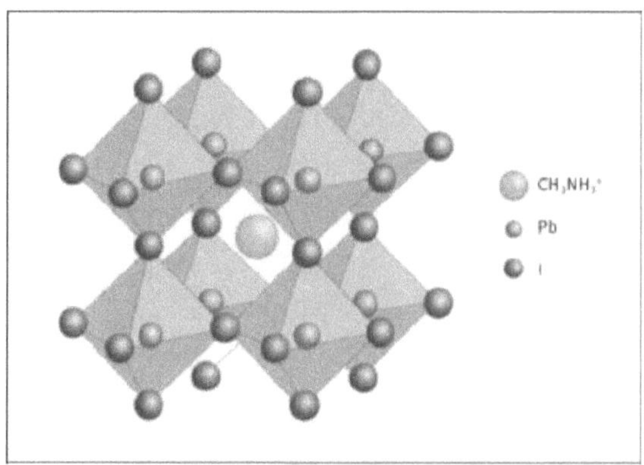

Figure 2.14. Crystal structure of $CH_3NH_3PbI_3$ (or MAPI) perovskite.[89]

ii)   *All-inorganic CsPbBr₃ perovskite properties and crystal structure*

$CsPbBr_3$ perovskite shares many of its properties with other perovskites, including direct bandgap, high light absorption coefficient, as well as easy and cost-effective solution-processed synthesis without the need for high-temperature preparation. However, $CsPbBr_3$ has other beneficial features, such as superior stability compared to other hybrid perovskites, especially at high temperatures.[90, 91] It is also characterized by higher luminescence intensity compared to $CH_3NH_3PbI_3$. For example, due to the narrower full width at half maximum (FWHM) of its emission peak, $CsPbBr_3$ can be utilized in high-definition displays.[92] Thus far, $CsPbBr_3$ perovskite has been widely used as an effective layer in different devices, as well as the light-sensing layer in photodetectors, due to its superior electronic properties, such as high mobility and ability to detect a broad spectrum of visible light, and its capacity to convert photons to photo-carriers with high responsivity.[93]

The crystalline structure of $CsPbBr_3$ at room temperature is tetragonally or monoclinically deformed. However, a pure cubic structure can be obtained at 130 °C.[91] Figure 2.15 shows the $CsPbBr_3$ perovskite 3D orthorhombic crystal structure (*Pnma*), with lattice parameters $a = 8.26$ A°, $b = 8.20$ A°, and $c = 11.75$ A°.[93]

In the work reported in this book (Chapter 6) MAPI perovskite was synthesized by Norah Alwadai, a Ph.D. student with Prof. Iman Roqan, whereas $CsPbBr_3$ perovskite was synthesized by Dr. Bin Xin, a postdoctoral fellow in Prof. Iman Roqan's group.

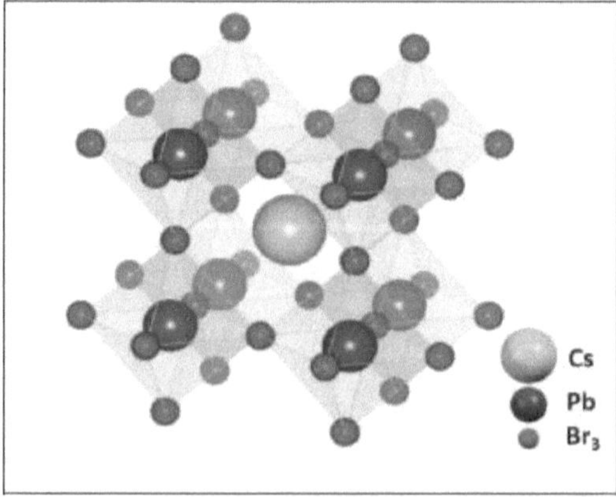

Figure 2.15. Crystal structure of inorganic $CsPbBr_3$ perovskite.[94]

2.9.2. Wide Band Gap MnO QDs

In the work reported in this book, highly crystalline p-type wide-bandgap manganese oxide-based QDs (p-MnO QDs) were used. MnO QDs are considered highly stable p-type wide bandgap semiconductors (WBGSs) ($E_g\sim$ 5 eV) operating in the DUV region.[95] These MnO QDs were synthesized by cost-effective solution-processed femtosecond laser ablation technique in liquid (FLAL), as described in Chapter 3. For the crystallographic and dimensional properties of these MnO QDs, the mean diameter was estimated at ~4.8 ± 0.2 nm and the interplanar spacing along the (200) and (111) planes was found to be 2.22 and 2.57 Å, respectively.[95] When the elemental composition of the p-MnO QDs was analyzed, the MnO phase was found to be dominant (81.5%), with smaller contributions of MnOOH (12.0%) and Mn_2O_3 (6.5%) phases.[95]

2.10. Optoelectronic Device Operational Mechanisms

Recently, GaN NWs have attracted significant research interest due to their potential use in a wide range of devices, such as LEDs, UV photodetectors, sensors, and transistors. Even though the work reported in this book focused primarily on the working principle of photodetectors, the LED working principle is briefly described, as the GaN NWs developed in this work can be potentially utilized in LED production.

2.10.1. Photodetectors

Photodetectors are devices capable of detecting light energy. The detected light is measured indirectly via measurements of photogenerated carrier current arising due to the photoelectric effect.[96, 97] When light is incident onto a semiconductor-based photodetector device, electron−hole pairs are generated, whereby carrier movement results in an electric current. When an external electric field is applied to the photodetector during illumination, photocurrent is obtained.[98] Owing to this property, the inorganic semiconductors, including III-nitrides, are widely used in photodetector applications. A simple photodetector device is depicted in Figure 2.16. Although GaN photodetectors are well-suited for UV-light detection applications, their performance in the visible range is inferior, suggesting the need for further improvements.[99] Therefore, extensive research has been conducted on GaN/perovskite devices to overcome this issue.

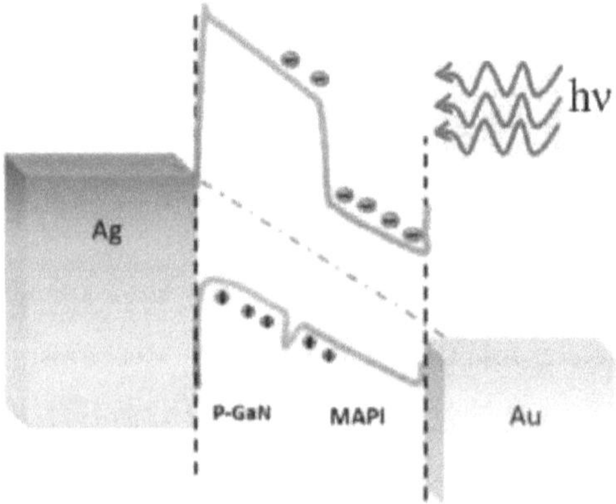

Figure 2.16. Schematic diagram of GaN/ CH₃NH₃PbX₃ perovskite-based photodetector.[100]

i) *Working principle of photodetectors and fundamental parameters*

When light falls on the photodetector surface and a positive/ negative bias to the anode (ITO)/ cathode (Au) is applied as shown in Figure 2.16, the photons will be absorbed by the material, forming an exciton. This exciton will be separated at the interface between the layers, creating electrons and holes (generated photocarriers). These electrons will travel along with the electron transport layer (in the work reported here, GaN NWs perform this function) to be collected at the cathode. The same principles apply to the holes, which are collected at the anode. [101] GaN/perovskite photodetectors can introduce a broadband photoresponse, where a thin layer of perovskite is enough for light detection in the visible spectral region.[102]

ii) *Photodetector parameters*

Some of the parameters characterizing photodetectors are discussed below:

a)   ***Dark current:*** Dark current is the static current produced within photodetectors in the absence of light exposure. In perovskite photodetectors, the main dark current source is the leakage current produced by carriers that are captured by the defect states inside the bulk material, as well as at the interfaces between perovskite and other material. However, dark current can also arise due to the thermionic emission resulting from heat (phonons) tunneling inside the material. As dark current is undesirable, considerable efforts have been made to minimize it, thus enhancing the photodetector performance.[103]

b)   ***Responsivity (R)***: When electromagnetic radiation is incident onto the photodetector, excitation of the charged photocarriers occurs, resulting in photocurrent. The R-value is introduced as a measure of the response efficiency of the photodetector to a light signal. Thus, R is defined as the ratio of generated photocurrent and incident light intensity and can be expressed as[104]:

$$R = \frac{\Delta I}{PS}; \qquad\qquad 2.20$$

where ΔI represents the difference between current produced under dark and illumination conditions, P is the incident power density, and S is the effective area illuminated by an external light source.

   c) ***Detectivity (D*)***: Detectivity allows estimating the minimum amount of light energy that a photodetector can detect. It is expressed as the ratio of R and the noise current density, and can be calculated using the expression below:[105]

$$D^* = \frac{R}{\sqrt{2qJ_d}} \; ; \qquad\qquad 2.21$$

where q represents the electron charge (1.6×10^{-19} C), and J_d denotes the dark current density.

   d) ***Response speed***: The response speed (or response time) of a PD device is measured through the rise time (t_{rise}) and the decay time (t_{decay}). The t_{rise} value is measured when the PD is switched on and indicates the time required to transition from 10% to 90% of the highest photocurrent value. Conversely, t_{decay} is measured when the PD is switched off. Fast response is indicative of high device performance and necessitates that the transient time exceeds the recombination time.[103]

2.10.2.   Light Emitting Diodes (LEDs)

LED is a device characterized by the operating mechanism opposite to that of a photodetector because LED emits light when current flows through it. In LEDs, the electrons and holes tend to diffuse in a direction determined by the applied forward field.[106] A simple LED structure comprises of an n-type layer, an active region where the recombination occurs, and a p-type layer, as depicted in Figure 2.17.[33] The carriers are injected by an external source through the metal contacts and move to the active regions, whereby radiative recombination occurs within the buffer GaN layer.

In the work reported in this book, PLD GaN NWs, buffer GaN (wetting) layer is serve as the n-type and active layer, and the p-doped GaN layer (obtained by doping with Mg) on sapphire substrate acts as the p-type layer, as well be shown in Chapter 5.

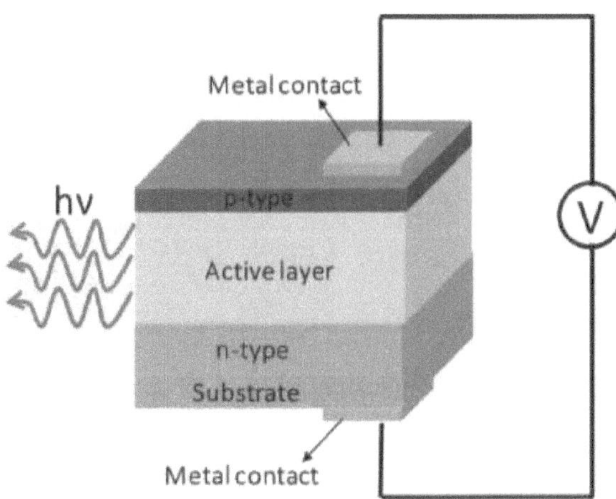

Figure 2. 17. Schematic diagram of a simple LED.

Chapter 3

Experimental Techniques

In this chapter, the experimental techniques that have been used for completing the work reported in this book will be briefly described. The chapter will commence with the growth and fabrication techniques, followed by the structural, optical, and electrical characterization techniques that were adopted to investigate the properties of the GaN NWs and related devices developed as a part of the present study.

3.1. Material Growth and Fabrication Techniques

3.1.1. Pulsed Laser Deposition Technique

Pulsed laser deposition (PLD) possesses several advantages relative to other deposition techniques, including simple operation and no requirement for toxic or expensive precursors. PLD is a simple and cost-effective physical vapor deposition technique that is suitable for preparing high-purity samples by stoichiometrically transferring the target material to the substrate.[107] Besides, PLD allows for easy sample handling, as the laser source is placed outside the reaction chamber.[108]

In a PLD system, a pulsed laser beam is applied to evaporate the material contained in the target and create a highly luminous plasma plume. Under controlled growth conditions, the ablated species escape from the target and condense upon arriving on the substrate surface, which is mounted on a heater at a particular distance above the target (see Figure 3.1).[109] To minimize particle scattering, the deposition procedure takes place in a vacuum chamber, which is equipped with three independent gas inlets, along with a flow controller, thus ensuring fixed pressure inside the chamber. The substrate temperature, the laser spot size on the target, and the distance between the substrate and the target determine the structure and uniformity of the material being grown. Furthermore, to avoid contamination, the target is pre-ablated prior to deposition.[110]

All GaN NW growth by PLD required for the present study has been performed by myself in the KAUST lab facility (Nanofabrication Core Lab).

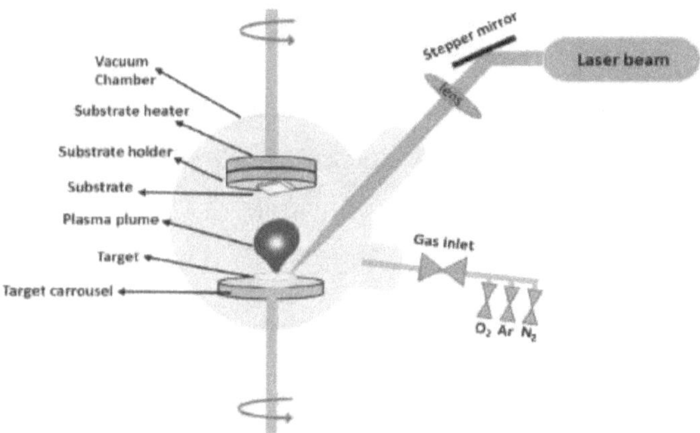

Figure 3.1. Schematic diagram of the PLD system used in the work reported in this book.

3.1.2. Femtosecond-Laser Ablation in Liquid

The femtosecond-laser ablation in liquid (FLAL) is considered a cost-effective material synthesis technique that can be used for different materials.[111] In this book, the FLAL technique was performed to synthesize high-quality solution-processed QDs. FLAL has many advantages, such as fast preparation, along with a simple and cost-effective synthesis procedure, and is carried out under atmospheric conditions, making it suitable for large-scale applications. In the FLAL setup used in the current work (depicted in Figure

3.2), a target immersed in a liquid is subjected to a femtosecond laser beam, which ablates the target, resulting in the formation of QDs within the liquid. As laser pulses do not interact with the material of the ablated species, the QD quality is maintained without any material damage, while avoiding the need for additional processes.[112] In the FLAL setup used in this study, the femtosecond laser ablation beam is focused on the target surface, while accounting for the amount of liquid above the target level, which is determined precisely, thus ensuring that the plasma plume is restricted to a small area due to which QDs are dispersed in the liquid solution. To obtain uniform QD distribution in liquid, the container holding the liquid and the submerged target was placed on a rotational stage, allowing the entire target surface to be exposed to the incident laser beam.[113]

In the study reported in this book, for the FLAL synthesis, titanium-sapphire femtosecond laser (Coherent Mira 900) located in Prof. Roqan's lab was utilized, at 800 nm wavelength. The MnO QDs were synthesized by myself and Dr. Somak Mitra, a postdoctoral fellow in Prof. Iman Roqan's group.

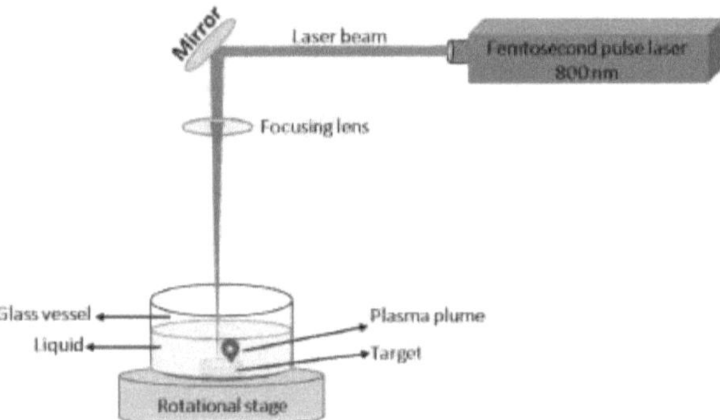

Figure 3.2. Schematic diagram of the FLAL setup adopted when performing the work reported in this book.

3.1.3 Magnetron Sputtering

Magnetron sputtering is one of the physical vapor deposition (PVD) techniques that are commonly used in industrial fabrication and device processing.[114] When applying this technique, a magnetic field intersecting the DC discharge electric field is used, confining the plasma to the area in front of the target, as shown in Figure 3.3. As the ionization rate increases, the deposition rate increases, resulting in shorter processing time. Due to the collisions between incident charged ions and the target surface, the target atoms are sputtered. During the sputtering process, the kinetic energy of incident charged ions is transferred to the surface atoms of the target, which travel across the vacuum chamber and condense on a substrate to form an epitaxial film. The quality of the grown layer is improved by reducing the pressure (maintaining it in the 10^{-1} Pa range), as this limits the likelihood of interaction between the gas species and the sputtered plasma.[115]

All sputtering deposition processes for fabricating indium tin oxide (ITO) contact layer and other metals electrodes (e.g., Au and Ag) for the devices described in this book were carried out by myself using ESCRD4 magnetron sputtering equipment produced by Support Company Ltd, available at the KAUST Core Lab (Nanofabrication Core Lab).

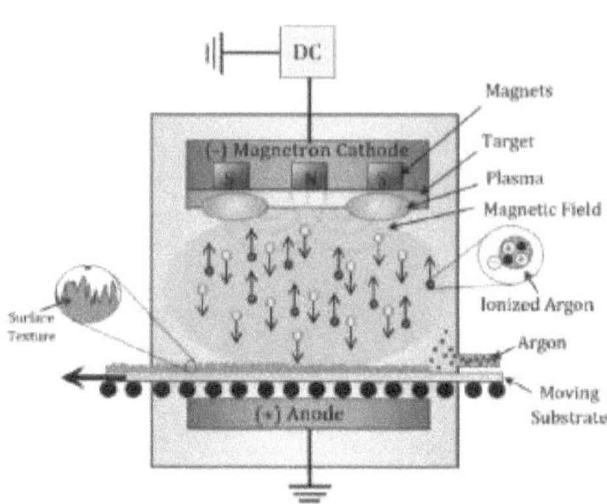

Figure 3.3. Schematic diagram of the magnetron sputtering setup used in the present work.[116]

3.2. Structural Characterization

3.2.1. X-ray Diffraction

X-ray diffraction (XRD) is one of the most important structural characterization techniques, which is typically used to investigate the crystalline properties, material elemental composition, lattice parameters, and growth direction.[117] The XRD setup includes a beam source, a stage on which the sample is mounted, and a detector, as depicted in Figure 3.4a. The diagram depicted in Figure 3.4b shows that when the X-ray beam with wavelength λ is incident on a sample at an angle θ and is diffracted from the material's crystal planes, constructive interference occurs when the conditions described by the Bragg's law are met:[117]

$$2n\lambda = 2d\,Sin\theta\ ; \qquad\qquad 3.1$$

where d is the lattice spacing between the planes and n is an integer number.

All θ−2θ measurements of all samples reported in this work were carried out by myself at the KAUST Core Lab using a Bruker D8 Advanced X-ray diffractometer.

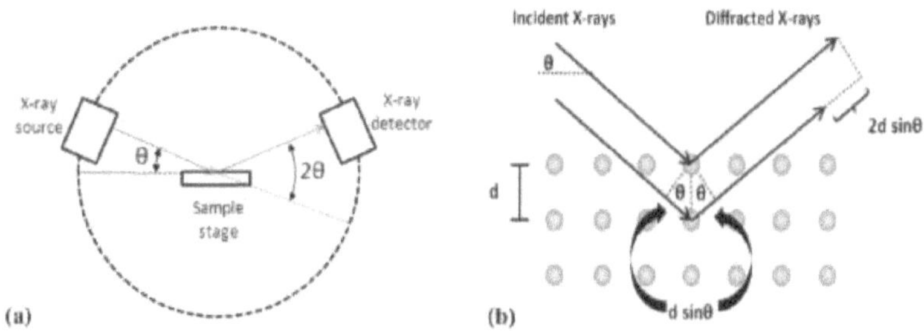

Figure 3.4. Schematic of (a) a typical XRD setup, and (b) the Bragg's law principle.[117]

3.2.2. Scanning Electron Microscopy

Scanning electron microscopy (SEM) technique helps in investigating material surface morphology.[79, 118] The SEM system employed in the study reported in this book is depicted in Figure 3.5a. For this purpose, an electron gun is heated to a high temperature to allow thermionic electron emission whereby escaping electrons are accelerated by a large electric field (1−10 kV). The accelerated electrons are focused by electromagnetic lenses, resulting in a spot of a small diameter (in the 1−100 nm range) on the sample surface. The electron beam (e-beam) incident on the specimen is reflected/scattered due to the beam−matter interaction, resulting in different generated signals and effects, as depicted in Figure 3.5b. To evaluate the sample morphology, secondary electron (SE), and backscattered electron (BE) signals are analyzed and are subsequently magnified produce to specimen surface images.[118]

To study the elemental composition of the specimen, energy dispersive X-ray (EDX) spectroscopy is typically utilized. If a SEM apparatus is adopted for this purpose, this necessitates a two-step process. In the first step, when the e-beam interacts with the sample, part of its energy is transferred to the sample's atoms. The electrons in these atoms are thus sufficiently energized to transition to a higher-energy shell or leave the atom. In both cases, the electron leaves behind a hole. In the second step, electrons from higher-energy shells transition to the lower-energy shell to fill the previously created hole, releasing energy equivalent to the difference between the energy levels of the shells involved in the form of an X-ray, as shown in Figure 3.6. As the energy difference depends on the atomic number, which is a unique property of every element, generated X-rays can be used to determine the elemental composition of the specimen.[119]

All SEM images required for investigating the surface morphology of samples examined in the present study were acquired by myself using FEI Nova Nano 630 SEM equipment at the KAUST Core Lab (Imaging and Characterization Core Lab).

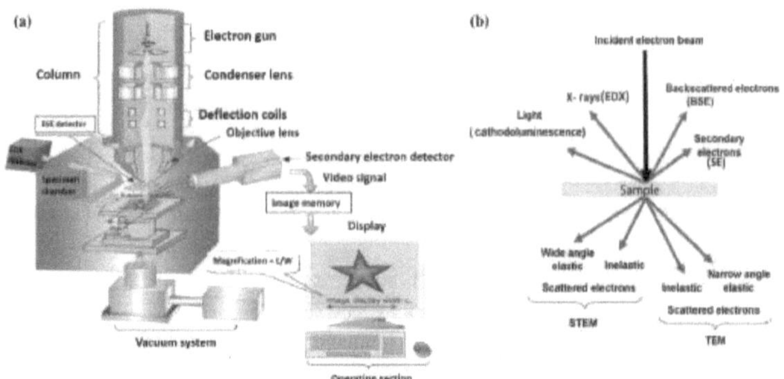

Figure 3.5. (a) Schematic diagram of a SEM setup, and (b) signals generated due to the e-beam interactions with the sample.[66, 120]

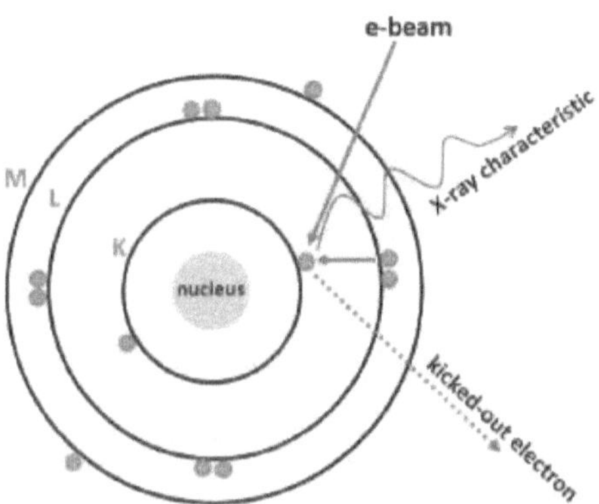

Figure 3.6. The fundamental principles of EDX spectroscopy.

3.2.3. Transmission Electron Microscopy

Transmission Electron Microscopy (TEM) allows atomic resolution imaging, which is employed for a more detailed characterization than can be achieved by SEM, due to its atomic-scale resolution. Therefore, it can be adopted to determine the crystal structure, the crystal orientation, crystal phases, structural defects, and the interface structures in the sample.[121]

The TEM operational principle is very similar to that of SEM. However, in TEM, instead of using the secondary electrons, as shown in Figure 3.5b, the e-beam transmitted through the specimen is recorded as an image captured by a charged-couple devices (CCD) array camera or is displayed on a fluorescent screen, as shown in Figure 3.7a. For the TEM technique to yield optimal results, the e-beam needs to be successfully transmitted through the specimen, which requires the electron acceleration voltages in the order of 10^5 V, and the specimen thickness below 100 nm. Thus, sample preparation for TEM measurements is a critical task that is carried out by a focused ion beam (FIB) in SEM. For high resolution-

TEM (HR-TEM), the resolution can reach 0.5 Å.

In addition to the aforementioned techniques, scanning-TEM (STEM) mode was also applied in this work, whereby a well-collimated and focused electron beam was employed for the transmission and scanning coils were applied to scan this beam. High-angle annular dark-field (HAADF) configuration was adopted for this purpose, which requires the use of a ring-shaped detector to detect electrons scattered when the main beam passes through the sample while ignoring the directly transmitted beam, as depicted in Figure 3.7b.

Moreover, electron energy loss spectroscopy (EELS) was used to identify the chemical composition of the studied samples. In EELS, the energy lost by the incident electrons during transitions is measured and compared with the characteristic ionization energy values of individual elements.

In this work, all HR-TEM and STEM measurements were performed by the KAUST Core Lab scientist, Dr. Sergei Lopatin, using Titan Themis Z (40-300) Thermo Fisher TEM, while the FIB sample preparations were performed by Ms. Nini Wei. All techniques are located at KAUST Imaging and Characterization Core Lab. The low-loss spectra were acquired in the so-called microprobe STEM mode with about 1 mrad semi-convergence angle (4 nm probe size).

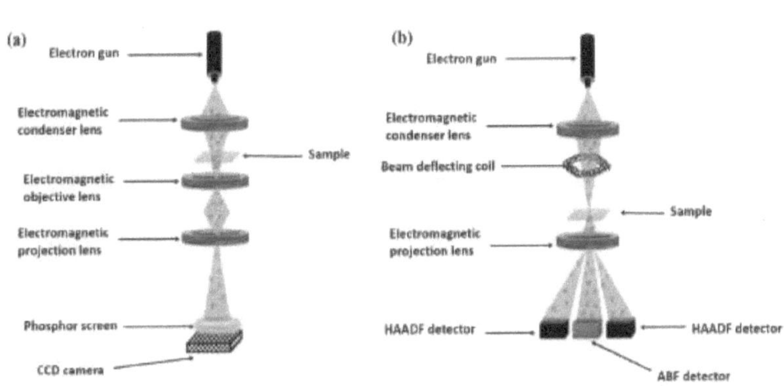

Figure 3.7. Schematic diagram of a typical (a) TEM setup and (b) STEM setup.[106]

3.3. Optical Characterizations

Several optical characterization techniques have been used in the work reported in this book to examine the optical properties of the studied materials, such as the bandgap of GaN NWs, and to explore the emitted spectra, aiming to identify the optical properties, as described below.

3.3.1. Photoluminescence Measurement

PL spectroscopy is a technique based on exciting materials by a laser beam with energy (wavelength), that is greater (shorter) than that of the semiconductor bandgap. As a result, free electrons in the CB and holes in the VB are created. The light emitted due to electron−hole recombination (as explained in Section 2.3.2) is dispersed by a grating located inside a monochromator (spectrograph) that is attached to a CCD camera,[37] as shown in the PL schematic setup depicted in Figure 3.8. A typical PL setup consists of a light source (laser or a monochromatic light source, such as Xe lamp attached to the excitation monochromator), lenses to focus the light on the sample, a sample holder, and finally the detecting monochromator that is attached to a detector, such as a CCD camera or photon multiplier tube (PMT). In this setup, a set of lenses is needed to focus the light emitted by the sample on the spectrograph slit.

In the work reported in this book, all temperature- and power-dependent PL measurements were performed by myself in the Semiconductor and Material Spectroscopy Laboratory run by Prof. Roqan, using He-Cd continuous wave (CW) with λ = 325 and Ar+ laser using second harmonic generating (SHG) crystal (producing 244 nm line), while room temperature (RT) PL was conducted by myself using Horiba Aramis spectrometer setup attached to He-Cd laser located at the Core Lab at KAUST (Physical Properties Characterization Core Lab). Temperature-dependent PL was performed at different temperatures (ranging from 5K to RT) while placing the samples in a helium-cooled cryostat.

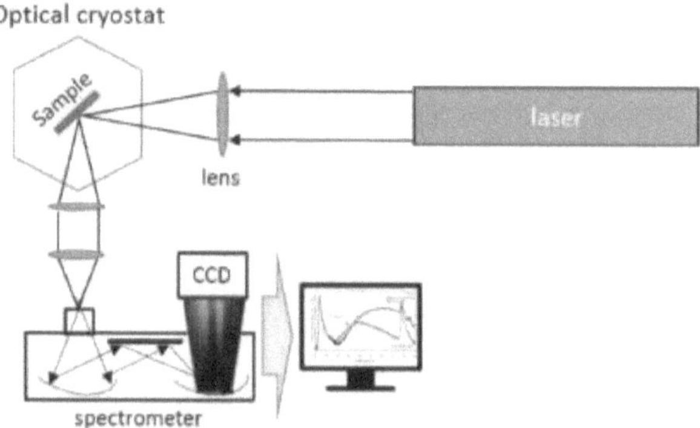

Figure 3.8. Schematic illustration of the PL setup adopted in the present study.[73]

3.3.2. Photoluminescence Excitation (PLE)

The photoluminescence excitation (PLE) technique is somewhat the inverse of PL, as it aims to identify the origin of emitted peaks and to measure the material bandgap. In addition, PLE measurements can be performed to determine material as well as quality, material bandgap and evaluate the emissions arising from the impurities or defects. A typical PLE setup contains the same basic elements as does a flexible tunable excitation source setup (lamp attached to excitation monochromator) while keeping the detected line fixed using emission monochromator.[26] As the PLE peaks correspond to the absorption edges of the studied material, they can be examined to investigate the excitation wavelengths of materials. Usually, in a PLE setup, a Xenon arc lamp attached to the excitation monochromator is employed, whereby its emitted band extends from deep UV (200 nm) to near-infrared (900 nm) part of the electromagnetic spectrum. A grating located in the monochromator is used to tune the excitation wavelengths. The excitation source bandwidth can be controlled by using a slit placed in front of the excitation monochromator, as depicted in Figure 3.9. The signal produced by the sample is acquired by an emission monochromator attached to a CCD or PMT.

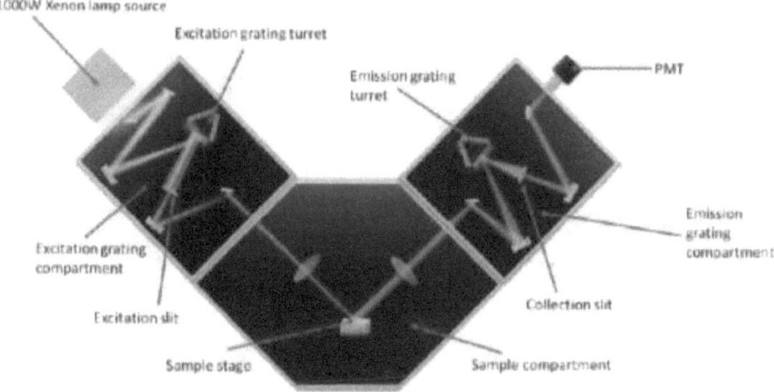

Figure 3.9. Schematic of a typical PLE setup.[106]

In the present study, Edinburgh Instruments FLS980 spectrometer attached to a 1000 W Xe lamp (Newport) available at the Prof. Roqan's lab was utilized by myself. This instrument consists of a PMT (R928P) detector with a $\lambda = 200-900$ nm detection range.

3.3.3. Time-Resolved Photoluminescence Spectroscopy

Time-resolved photoluminescence (TRPL) spectroscopy is one of the most important optical characterization techniques for investigating the carrier dynamics of materials. In particular, TRPL is essential for gaining an understanding of the exciton lifetime of radiative and non-radiative processes, as these parameters are considered the key indicators for material/device quality.[122] In a typical TRPL setup, ultrafast pulsed laser serves as an excitation source, allowing the PL spectrum to be obtained at different ultrashort time intervals (for determining the carrier decay times), allowing the PL spectrum to be obtained at different time points, while a streak camera can be used to capture the detected events. In the study presented in this work, titanium: sapphire (Ti: sapphire) ultrafast pulsed femtosecond (fs) laser system (Mira 900D from Coherent) was employed and was pumped by Coherent diode-pumped solid-state CW laser (Verdi DPSS with $\lambda = 532$ nm). The third harmonic generator (THG) supplied extra wavelength tuning capability (extending it to the UV range) and was used to excite the GaN samples by energy equivalent to $\lambda = 266$ nm. Also, for temperature-dependent TRPL measurements, the sample was held in a helium-cooled cryostat. Finally, to study and detect ultrafast phenomena characterizing the light emitted by the samples, a streak camera (Hamamatsu C6860) was employed. To perform TRPL, two streak camera operational modes can be used depending on the decay time of carriers in each material. These modes are single-sweep mode (for a long decay time, which extends from hundreds of ps to the ms range) and syncroscan (for ultrafast decay of $1-2$ ps duration). Figure 3.10 shows the TRPL setup adopted in the present study. The emitted spectra were collected by the Princeton Instruments spectrograph connected to the streak camera. When the emitted light reached the streak camera slit, it was focused into the photocathode of the streak tube, after which the photon signals were converted to several electrons that were accelerated through a chamber before striking a microchannel plate (MCP). As a result, these photoelectrons could be converted by MCP to a live decay image displayed on the phosphorus screen.[73]

All TRPL measurements required for the work presented in this book were performed by Dr. Somak Mitra, a postdoctoral fellow in Prof. Roqan's group, in the Semiconductor and Material Spectroscopy Laboratory run by Prof. Roqan.

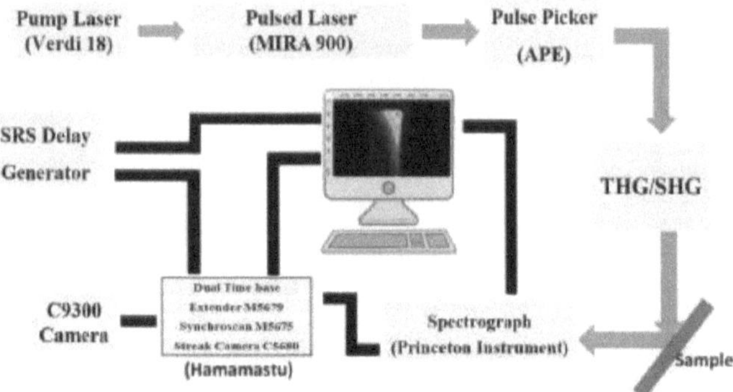

Figure 3.10. A schematic of the TRPL setup used in this study.

3.3.4. UV-VIS Absorption Measurements

UV-VIS absorption technique is adopted to evaluate the bandgap of materials (GaN NWs in this work), as well as to explore the absorption spectra produced by the material. A typical UV-VIS setup comprises a light source, a sample holder, and a monochromator attached to a detector, which is used to acquire the transmitted (reflected) spectra generated by the sample (using transmission mode and reflection mode for transparent and non-transparent samples, respectively). A frequency sweep is produced by a monochromator, allowing full system response to be investigated, as shown in Figure 3.11a.[26]

For this work, absorption measurements were performed by myself in the Analytical Chemistry Lab at the KAUST Core Lab by using Varian Cary 5000 UV-VIS-NIR spectrophotometer, as shown in Figure 3.11b.

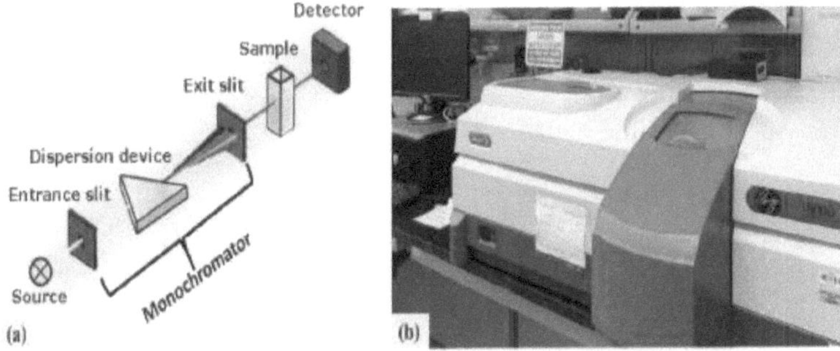

Figure 3.11. (a) Schematic diagram of Varian Cary 5000 UV-Vis-NIR spectrometer setup, and (b) Varian Cary 5000 UV-Vis-NIR employed in this work.

3.4. Complementary Techniques Used by Collaborators

3.4.1. Cathodoluminescence

Cathodoluminescence (CL) is occurred by e-beam-matter interaction. The incident e-beam on the sample generates CL spectra that are detected, as shown in Figure 3.5b. The CL spectra are dispersed by a monochromator and are detected with a CCD camera. A CL system can be attached to a SEM apparatus, thus benefitting from the very high SEM resolution (down to 1 nm). As a result, a full CL spectrum can be obtained from each pixel in the CL map because the wavelength of the electron is in an angstrom range. Besides, the optical properties of a sample can be correlated to structural features observed by SEM.[123] A typical CL setup is shown in Figure 3.12.

In the work reported in this book, for RT cathodoluminescence (CL) hyperspectral imaging, a custom-built acquisition system in an SEM can be adopted (5 kV energy and 100 ms/pixel acquisition time was chosen for this study). All CL measurements were performed by Dr. Paul Edwards, a researcher working in the Spectroscopy and

Semiconductor Devices group led by Prof. Robert Martin in the Physics Department, University of Strathclyde, Glasgow, United Kingdom.

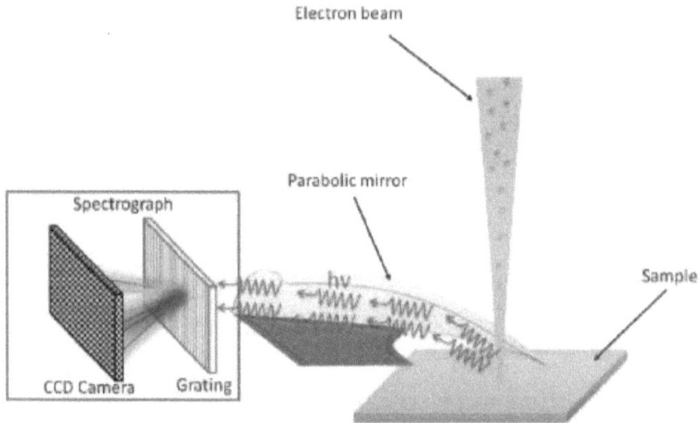

Figure 3.12. Schematic diagram of a typical CL setup.[106]

3.4.2. Focused Ion Beam (FIB)

Focused ion beam (FIB)-SEM is widely used to prepare TEM samples. FIB has the same operating principle as SEM, with the e-beam being replaced by an ion beam.[124] These ion beams can be used for imaging, material deposition, or milling. Initially, a thin carbon film is deposited, followed by a thick platinum (Pt) film to protect the sample surface. Ga^+ ions are used as ions beam in FIB. Once a thin part of the sample surface ($\sim15\times5$ μm^2) is selected, the ion beam mills the chosen area to create a trench-like shape. The thin tablet-like cross-section is cut from the sample carefully and attached to an Omni-probe located inside the FIB by using Pt. The thus prepared material is transferred to a grid and polished carefully by vertical milling to obtain the required TEM dimensions.[124]

All TEM lamellas used in this book were prepared by the KAUST Core Lab (Imaging and Characterization Core Lab) staff Nini Wei and Alessandro Genovese using FEI Helios FIB-SEM.

3.4.3. Molecular Beam Epitaxy (MBE

Molecular beam epitaxy (MBE) is a widely used technique for growing different high-quality semiconductor structures (thin films or nanostructures) on different substrates.[125] The MBE system consists of three main components, i.e., the growth chamber, the preparation chamber, and the fast entry lock (load-lock chamber).[126] Figure 3.13a and 3.13b respectively depict the schematic diagram and the actual MBE technique system. Each of the three chambers is separately connected to different pumps and is isolated from others by gate valves, allowing the vacuum level to be maintained in each chamber separately from others. The samples are loaded into a load-lock chamber on a molybdenum block holder. The pre-outgassing step is executed in the preparation chamber to protect the growth chamber from water and other contaminants. Finally, samples are placed into the growth chamber to grow some layer or another structure. Stainless steel is used to construct both the growth chamber and the gate valves that are used to isolate it.

In the work reported in this book, VEECO GEN 930 Plasma-Assisted (PA) MBE in the Photonics Laboratory run by Prof. Boon S. Ooi was used to grow InGaN/GaN MQWs on PLD GaN NWs. Dr. Jung-Wook Min, a research scientist in the Photonics Laboratory group performed all PA-MBE growth processes needed to obtain InGaN/GaN MQWs/PLD GaN NWs.

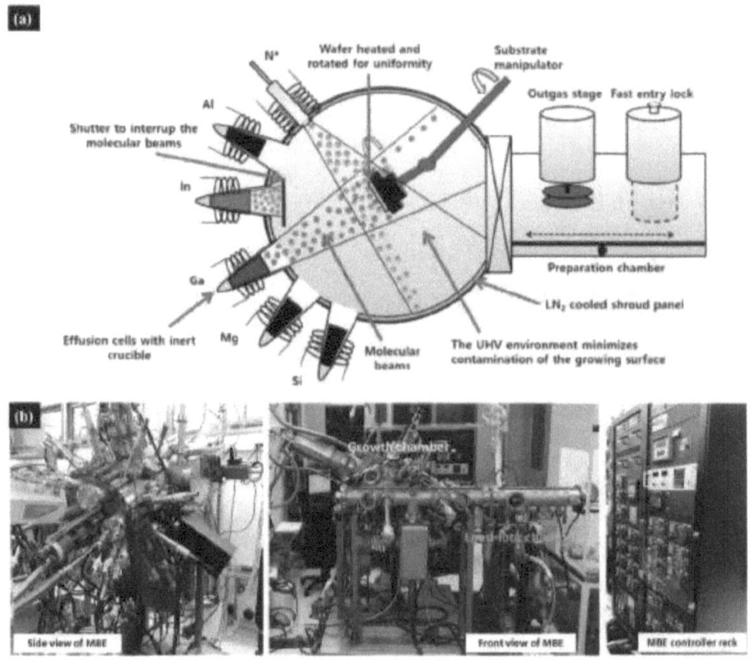

Figure 3.13. (a) A schematic diagram of a typical PA-MBE setup, and (b) VEECO GEN 930 PA-MBE employed in this work.

Chapter 4

High-quality self-assembly GaN nanowires (NWs) grown on different substrates by using pulsed laser deposition

4.1. Introduction

III-nitride semiconductors, mainly GaN-based materials, have inherent outstanding properties for a wide variety of optoelectronic, high power and high-frequency electronic applications such as light-emitting diodes (LEDs) and laser diodes (LDs), as well as devices used for sensing, medical curing, and industrial applications in harsh environments.[18-22] This is due to the unique properties of these materials, such as excellent thermal stability, their direct and tuneable bandgap, mechanical durability.[15, 17] However, direct growth of GaN with sufficient quality on any affordable or fixable substrate, which is the practical goal that the industrial market needs to achieve, is still challenging due to several issues. One of the most important issues that are the lattice mismatch between the substrates and III-nitride materials (e.g., the lattice mismatch between GaN and "Si and Al_2O_3" substrates is 17% and 14%, respectively[127, 128]) induces threading dislocation (TD) defects at the GaN interface with substrates, reducing the optical and electrical device performance. As a result, GaN-based devices still suffer from several limitations, including lower external efficiency ($\leq 10\%$) in the UV spectral range ($\lambda < 365$ nm).[22] In this case, the high TD density, generates a high density of nonradiative centers, markedly reducing the internal quantum efficiency (IQE).[22] While a native GaN substrate can eliminate this problem, this approach is extremely expensive.[22-15, 17] Also, although high quality of GaN grown on common substrates such as Si, SiC, and sapphire has been demonstrated, this design introduces other issues (e.g., the different thermal expansion of sapphire, the high absorption of Si for UV emission, and the high cost and size limitation of SiC substrates).[15, 17] Thus, rapid technological advances, and the emergence of potential GaN-based applications, such as flexible displays, bio-sensors, vertical UV and deep UV LEDs, and vertical-cavity surface-emitting lasers (VCSELs), increase the need for high-quality GaN growth on a wide range of substrates. However, the lift-off process and flip-chip fabrication used to transfer the device to the desired substrate further increase the design complexity and cost of production,[129] while causing structural damage, which severely limits the size and throughput of the released GaN-based devices, remaining a serious issue.[130]

A wide range of efforts has been reported to overcome these limitations.[22] The use of III-nitride nanowires (NWs) was identified as a potential solution to mitigate the substrate and high TD density issues.[131-136] In other studies, two-dimensional (2D) substrates have been used as a sacrificial layer for lifting-off devices, and several growth methods have thus far relied on 2D materials such as graphene,[137] boron nitride, and some transition-metal dichalcogenide (TMDs) substrates,[138] which have better potential compared to graphene for GaN growth due to the scalability limitation of graphene.[138] On the other hand, conductive or patterned substrates[131-136] are preferred for vertical high-quality GaN NW growth, limiting the range of suitable substrates. For example, recently, Ga_2O_3 was reported as one of the most competitive substrates for vertical UV and deep UV III-nitride devices as it is conductive and transparent.[56, 139] However, no GaN NW growth has been achieved on the Ga_2O_3 substrate. Likewise, direct growth of single-crystal GaN has been achieved on a very limited number of polycrystalline substrates, such as GaN/AlGaN epilayer on diamond (which is limited for layer structure growth),[140] III-nitride epilayers on polycrystalline metal substrates,[141] or using an interlayer between GaN and the substrate (e.g. MXene, graphene,[142, 143] or Ti on glass)[143, 144] to obtain the single-crystal GaN nanostructures.[144] Furthermore, although the density of TDs that propagate into the NWs is reduced extensively, the TDs in the interface between the material and substrate are not eliminated due to the lattice mismatch issue.

As the growth of modern technology requires different types of promising and suitable substrates for emerging applications such as bio-sensors, the aforementioned issues must be overcome.[22, 141-143] This can be achieved via the formation of a pre-polycrystalline GaN layer on any substrate to disable the TD growth in the interface between grown materials and substrates and assist the direct growth of single-crystal GaN on any substrate. This strategy is proposed, as different grain orientations in a polycrystalline material prevent the lattice mismatch (crystallographic relation) between the substrate and grown material.[145, 146] However, such a strategy has never been employed for growing high-quality single-crystal GaN via polycrystalline GaN on any substrate that can be used as a template to fabricate III-nitride-based devices.

In this chapter, we demonstrate a novel, universally applicable, and cost-effective strategy for growing self-assembled vertical GaN NWs directly on a wide range of available bulk and 2D substrates irrespective of the substrate type and lattice mismatch, using pulsed laser deposition (PLD) without a metal catalyst. The superior quality of these single-crystal NWs confirms that this growth strategy can be generalized for a wide range of large-scale applications, including flexible devices.

4.2. Experimental method

4.2.1. Structural Characterizations

Structural properties were carried out by SEM (FEI Nova Nano 630), and high-resolution XRD (Bruker D8 Discover with Cu K_α and $\lambda = 1.5406$ Å) methods.

4.2.2. Optical properties

Material optical properties were examined by PL measurements at room temperature (RT) by using a 325 nm continuous-wave (CW) He–Cd laser, while an Andor spectrograph connected to a charge-coupled device camera was used to collect the data. RT Raman spectroscopy measurements were applied by using Horiba LabRAM Aramis with continuous-wave Cobolt Blues 25™ laser ($\lambda = 473$ nm). RT absorption measurements were carried out using UV-vis Varian Cary 5000 spectrophotometer, aiming to identify any absorption edges that could be attributed to our samples.

4.3. Results and discussions

4.3.1. Substrates used in this work

In this work, we used two kinds of substrates bulk and 2D substrates. For bulk substrates, we used single-side-polished p-type Si (100) substrate with very low resistivity (0.001–0.005 Ω cm), p-type GaN layer doped by Mg (0.5 μm) grown on the c-sapphire wafer (1×1 cm^2) (University Wafer Inc.), a commercial bulk Ga_2O_3 from novel crystal (the details of the Ga_2O_3 growth discussed in other where),[147] and one-side-polished c-plane sapphire wafer (Semiconductor Wafer Inc.). For 2D substrates, we used graphene substrates (single-layer CVD graphene/90 nm SiO_2/highly (100) p-doped Si (wafer, Alpha

Graphene Inc.), Ti_3C_2 MXene, with ~ 40 μm particle size, (from Sigma Aldrich) deposited on glass by spray-coating and 2D TMDs (MoS_2 and WSe_2) substrates. All substrates were commercials apart from 2D TMD substrates. For 2D TMDC substrates, high-quality monolayer TMDs were synthesized on sapphire substrates by a typical chemical vapor deposition (CVD) method in the hot-wall furnace. High-purity WO_3 (Sigma-Aldrich, 99.9%), MoO_3 (Sigma-Aldrich, 99.5%), Se (Sigma-Aldrich, 99.99%), and S (Sigma-Aldrich, 99.5%) powders were used as precursors. The growth temperatures for WSe_2 and MoS_2 were 900 and 800°C, respectively. Ar and Ar/H_2 flowing gas was used as a carrier gas for MoS_2 and WSe_2 growth, respectively. All the growths were performed at low pressure (10-40 Torr). All 2D TMDC substrates were fabricated in Prof. Vincent Tung lab by his Ph.D. student Yi Wan.

For bulk substrates, before commencing PLD growth, the substrates were ultrasonically cleaned in acetone for 20 minutes, after which they were rinsed with deionized water for 20 minutes, repeating this process twice.

4.3.2. High-quality GaN NWs growth

We recently demonstrated that catalyst-free high-quality ZnO NWs and NTs can be grown on different substrates by precisely controlling the PLD conditions, which is considered relatively cost-effective.[108, 148, 149] However, no high-quality GaN materials were grown using previous PLD methods.[69, 150, 151]

Pioneer 240 PLD system (Neocera) equipped with a KrF excimer laser (of 248 nm wavelength) was used to ablate a commercial GaN target (99.95% purity) of 2.5 cm diameter. Different substrates were placed in the chamber at the same time. The target was loaded into the chamber and was secured to a rotating target holder to ensure uniform ablation and avoid drilling through the target by the continuous irradiation of the laser of intense laser pulses. To grow GaN NWs, the target was exposed to the laser beam to ablate the species towards substrates. The following optimized conditions were utilized, as they were experimentally determined to be the most optimal (as discussed later in this work): 150–250 mTorr nitrogen pressure (P_{N2}); 1 ± 0.05 J·cm^{-2} beam fluence focused on the GaN target surface (with 40,000–100,000 laser pulses delivered at a frequency of 10 Hz); ~9 cm vertical distance between the substrate and the target; and 850 °C growth temperature.

By optimizing the PLD conditions (as described in Section 4.3.3), we have achieved a novel and highly successful growth strategy that can be applied for obtaining vertically-aligned self-assembly GaN NWs grown directly on a wide range of bulk (Si, GaN, sapphire and (-201) β-Ga$_2$O$_3$) as well as 2D (MXene, graphene, MoS$_2$, and WSe$_2$) substrates, regardless of the lattice mismatch between GaN and the substrates and the substrate type, crystallinity, or orientation. It is worth mentioning that this the first attempt to grow GaN NWs on β-Ga$_2$O$_3$ substrate. The schematic diagram of our PLD growth mechanism adopted in this work is shown in Figure 4.1.

Figure 4.1. The schematic diagram of the PLD growth mechanism adopted in this work (Illustration created by Ivan Gromicho. Scientific Illustrator, Research Communication and Publication Services. Office of the Vice President for Research. King Abdullah University of Science and Technology (KAUST)).

4.3.3. Optimizing PLD conditions

To optimize the results, we have examined the effect of growth conditions.

i)   *Substrate temperature effect on GaN NW growth*

One of the factors that affect GaN NW growth is temperature.[152] Available evidence indicates that increasing the growth temperature is effective in improving the crystallinity, composition, and purity of growth.[152, 153] Moreover, it has been experimentally determined that to produce GaN single-crystal structure, high temperatures (700–1400 °C) and/or high nitrogen pressures are necessary to overcome the large kinetic barrier to crystal formation.[153] Consequently, in the present study, the maximum temperature permitted by the PLD system (850 °C) was adopted, as it was also previously used in the MBE system.[132]

It is also noteworthy that, below this value (i.e., at 800 °C and 825 °C) NWs could not be obtained, as shown in Figure 4.2. Figure 4.3 indicates the corresponding PL spectra taken from the samples shown in Figure 4.2, confirming an intense NBE emission from the optimized sample, whereas the other (not optimized) samples show low optical quality.

Figure 4.2. SEM images of GaN NWs growth at (a) 800 °C, (b) 825 °C, and (c) 850 °C.

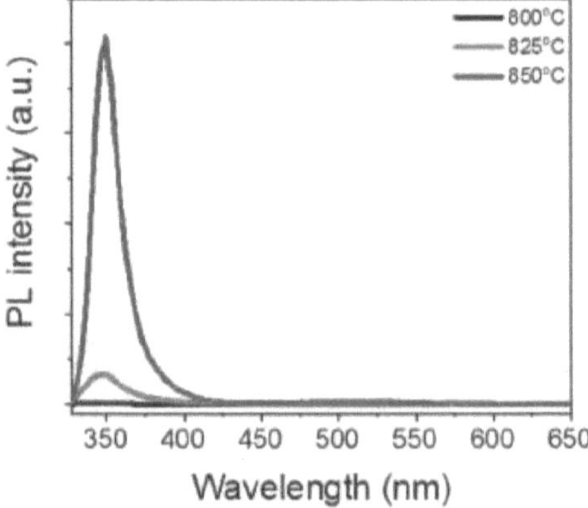

Figure 4. 3. The corresponding PL spectra of GaN samples grown at different temperatures.

ii)   *Laser power effect on GaN NW growth*

The energy of species (plasma) landed on the substrate (E) was also optimized by controlling the laser energy (i.e., power) as explained in Section 2.6.2. To optimize the laser power, the values of all other parameters except laser energy (which was measured inside the PLD chamber by using a power-meter) were kept constant. The laser power was measured on the target surface, ranging from 50 mJ (0.5 J/cm^2) to 120 mJ (0.7 J/cm^2). However, laser energy in the 50−70 mJ (0.5–0.7 J/cm^2) range failed to produce NW structures. At 80 mJ (corresponding to 0.8 J/cm^2 energy density on the target), initiation of NW structures was finally observed. Further improvements in the NW structure shape were attained at 90 mJ (0.9 J/cm^2), as well as at 100 mJ (1 J/cm^2). The findings revealed that, when the adopted energy exceeds the optimized E value that produces NWs, which can be achieved by reducing nitrogen pressure (P_{N2}) or increasing laser fluence on the target, a film structure is formed, as the ablated species in the emitted plume arrive at the substrate with very high kinetic energy E, as shown in Figure 4.3. At 1.2 J/cm^2 (120 mJ), no NWs were observed, and the structure reverted to the granular film form, as shown in Figure 4.4. Figure 4.5 indicates the corresponding PL spectra taken from the samples shown in Figure 4.4, confirming an intense NBE emission from the optimized sample, whereas the other (not optimized) samples show low optical quality.

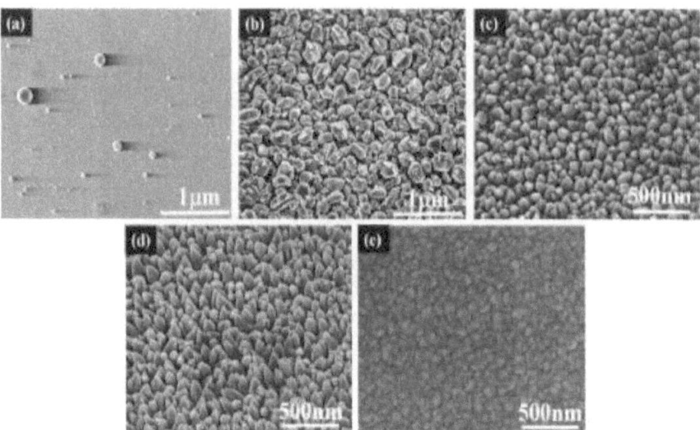

Figure 4.4. SEM images of GaN growth at (a) 0.6–0.7 J/cm^2 (60–70 mJ), (b) 0.8 J/cm^2 (80 mJ), (c) 0.9 J/cm^2 (90mJ), (d) 0.95–1.05 J/cm^2 (95–105mJ) and (e) at 1.2 mJ/cm^2 (120 mJ).

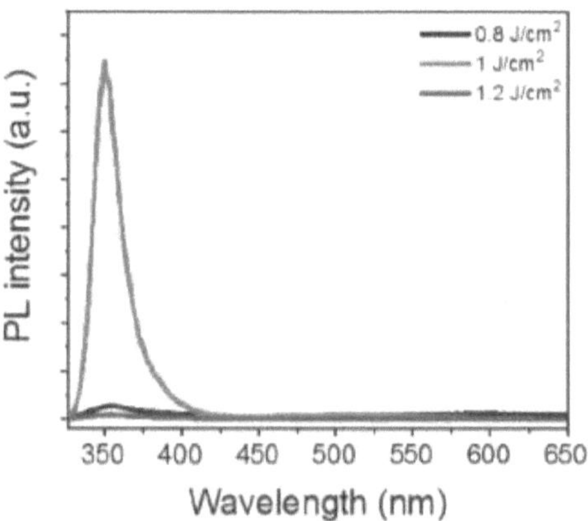

Figure 4. 5. The corresponding PL spectra of GaN samples grown at different fluences.

iii)   *Nitrogen pressure effect on GaN NW growth*

Available evidence indicates that, as the pressure increases, the energy of ablated species declines as a result of collisions with gas molecules, due to which the ablated species, once condensed, form nanoparticles.[149, 154] During GaN growth by PLD, the collisions occurring at high pressure can reduce the energy of the ablated species, allowing NW formation and preventing the epitaxial film growth.[155] In this work, as shown in Figure 4.6a and 4.6b, NWs were not obtained below 150 mTorr, but GaN films were attained instead (Figure 4.6). In order to determine the optimal GaN NW growth conditions when applying the PLD technique, as a part of this investigation, growth pressure was optimized by changing the P_{N2} value from 150 to 200 and finally to 250 mTorr (Figure 4.6c and 4.6d). Figure 4.7 indicates the corresponding PL spectra taken from the samples shown in Figure 4.6, confirming an intense NBE emission from the optimized sample, whereas the other (not optimized) samples show low optical quality.

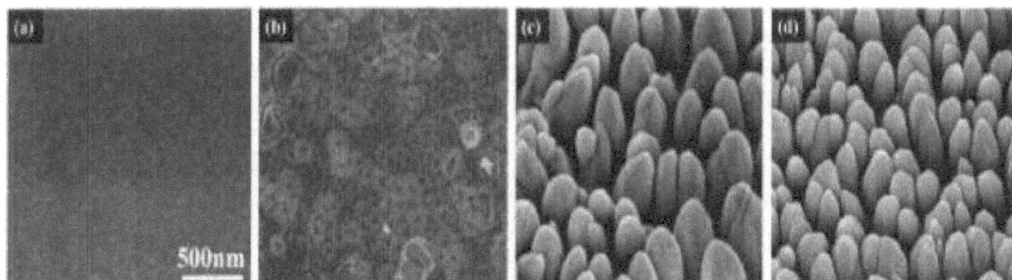

Figure 4.6. SEM images of GaN growth by PLD at (a) 50 mTorr, (b) 100 mTorr, (c) 150 mTorr and (d) 200 mTorr.

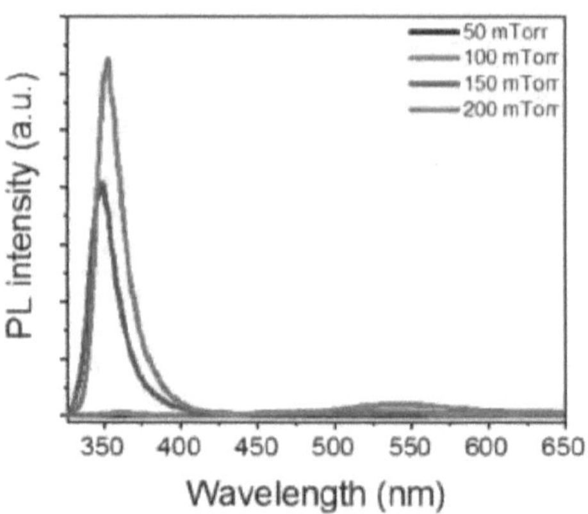

Figure 4.7. The corresponding PL spectra of GaN samples grown at different pressures.

iv) *The influence of the number of laser pulses on GaN NW growth*

As a part of the current study, it was determined that NWs could not be produced using 30,000 laser pulses when other PLD parameters are set to their optimized values, as shown in Figure 4.8a. Thus, the 40,000 − 100,000 pulse range was examined more closely, and the results yielded are shown in Figure 4.8b–4.8f. Specifically, in the 40,000–60,000 range, GaN NW length varied from 1.5 μm to 1.75 μm, as shown in Figure 4.9a, increasing to 3.3 μm when the number of pulses was in the 80,000–100,000 range, as shown in Figure 4.9b. It is also noteworthy that the diameter increases with the NW length. For example, for NWs of 1.5 μm length, diameters ranging from ~100 to ~200 nm were measured, depending on the substrate. Figure 4.10 indicates the corresponding PL spectra taken from the samples shown in Figure 4.8, confirming an intense NBE emission from the optimized sample, whereas the other (not optimized) samples show low optical quality.

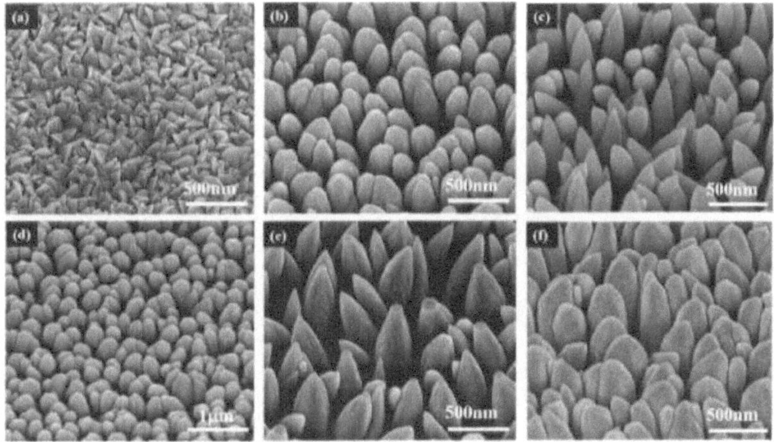

Figure 4.8. SEM images of GaN growth at (a) 30,000 pulses, (b) 40,000 pulses, (c) 50,000 pulses, (d) 60,000 pulses, (e) 80,000 pulses and (f) 100,000 pulses.

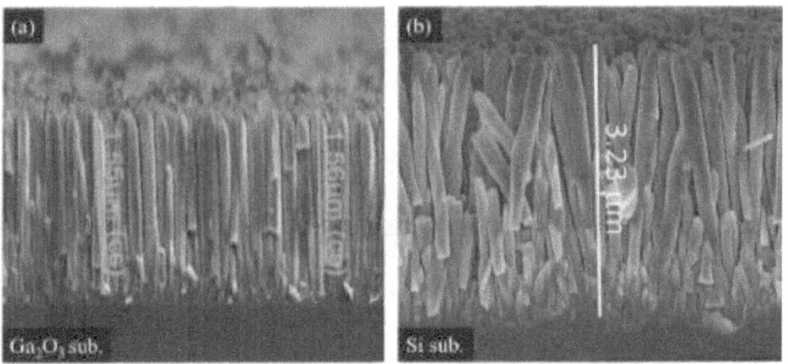

Figure 4.9. Cross-sectional SEM images of GaN NWs under optimal conditions, using (a) 40,000–60,000 and (b) 80,000–100,000 laser pulses.

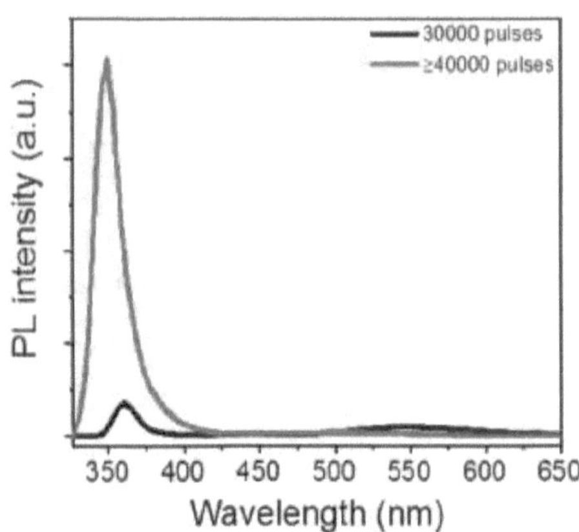

Figure 4. 10. The corresponding PL spectra of GaN samples grown using a different laser.

v) ***The effect of substrate type (bulk and 2D structures) on GaN NW growth***

Finally, the effect of different substrates (while varying every growth parameter) on the shape of the obtained structure (NW or film) was investigated. The optimal conditions for GaN NW growth on all substrates without catalysis or seeding are reported in Table 4.1. For the first time, the newly PLD strategy was developed to grow, vertically-aligned self-assembly GaN NWs grown simultaneously in the same PLD chamber on different bulk (Si, GaN, Ga_2O_3, and sapphire) substrates were obtained, as indicated by the SEM images presented in Figure 4.11a–4.11d. Besides, by using similar optimized PLD conditions, NWs were grown using 2D substrates (MXene, graphene, MoS_2, and WSe_2), as shown in Figure 4.12a–4.12d, regardless of the lattice mismatch.

Table 4. 1. Optimal GaN deposition conditions.

Substrate Temperature	850 °C	Laser Energy	95–105 mJ

Distance between target and substrate	9 cm	Pulse Duration	40000–100000 pulses
Pressure	150, 200 and 250 mTorr	Pre-ablation pulses	1000 pulses

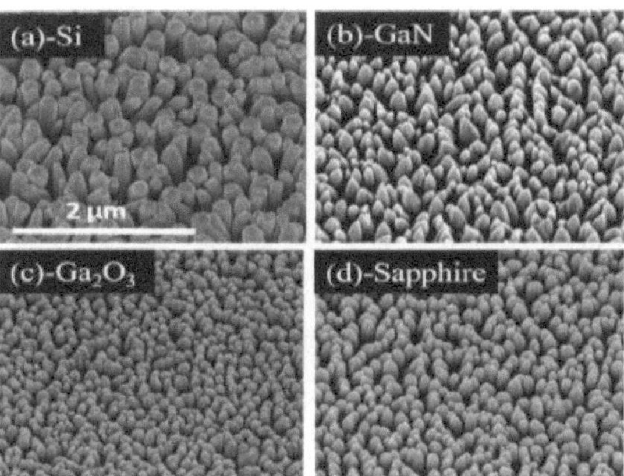

Figure 4.11. Tilted-view SEM images of self-assembly GaN NWs grown simultaneously in the same PLD chamber on (a) Si, (b) GaN (c) Ga_2O_3, and (d) sapphire substrates.

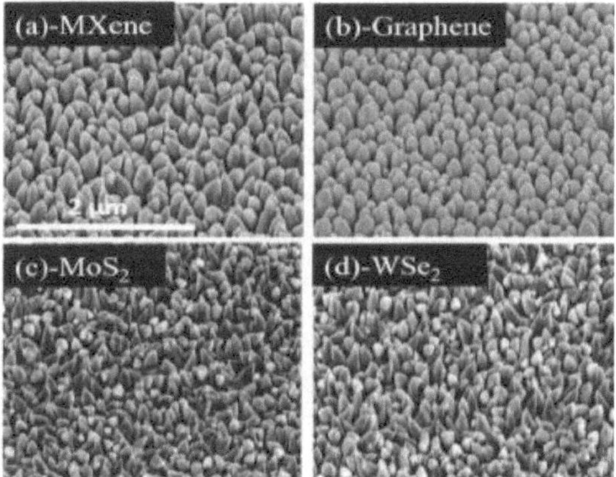

Figure 4.12. Tilted-view SEM images of self-assembly GaN NWs prepared by PLD technique on (a) MXene, (b) graphene, (c) MoS2, and (d) WSe2 substrates.

4.3.4. Characterizations

i)   *Structural properties*

The quality of GaN NWs grown on the 2D and bulk substrates was investigated via X-ray powder diffraction (XRD) and Raman measurements. Figure 4.13 reveals the crystalline structure of GaN NWs grown on all substrates. These XRD spectra show GaN peaks at 32.6°, 34.6°, 36.9°, 57.8°, and 63.5°, which are assigned to the (100), (002), (101), (110), and (103) planes of GaN, respectively, thus confirming the hexagonal wurtzite structure (ICDD 50-0792).[156, 157] However for all samples, (002) is the dominant peak and originates from the single-crystal NWs, whereas the other peaks are produced by the GaN polycrystalline wetting layer (WL) formed between the NWs and substrates, as will be shown later (Chapter 5 Section 5.3.1) using high-resolution (HR) STEM. The sharp (002) GaN peak, irrespective of the substrate employed, confirms high crystal quality. On the other hand, the XRD peaks at ~69° and 41.6° are ascribed to Si (400) and c-sapphire (006) planes (ICDD 27–1402 and ICDD 43–1484), respectively, while peaks at 18.8, 38.3, and 59.02° were attributed to the β-Ga_2O_3 (-2.1)[128] plane and its second and third-order, and the peak at 28.8° is assigned to the MXene (008) plane.[158, 159 160]

As the graphene, MoS_2, and WSe_2 monolayers are extremely thin to be detected by XRD, the corresponding peaks cannot be discerned from the XRD spectra. we used micro-Raman measurements for these samples using a 473 nm excitation line. Figure 4.14 shows the Raman spectra of GaN NWs grown on the 2D substrates, revealing that typical Raman peaks corresponding to GaN (located at 258, 421, 566, and 724 cm^{-1})[161] are obtained with a slight shift compared to bulk due to strain experienced by NW structure,[149, 160] which also depends on the NW diameter. In particular, the Raman peak located at 566 cm^{-1} is ascribed to the E_{high}^2 mode of wurtzite GaN.[160, 161] Graphene Raman peaks are observed at 1366 cm^{-1} (D) and 1600 cm^{-1} (G),[162, 163] while MXene peaks are observed at 1386 and 1590 cm^{-1}.[164] For NWs grown the TMD substrates, the MoS$_2$ Raman peaks at 383 cm^{-1} (E$_{2g}$) and 404 cm^{-1} (A$_{1g}$) are shown,[165, 166] while WSe$_2$ (A$_{1g}$) peak is shown at ~ 246.5 cm^{-1}.[166, 167] No significant differences are observed in the Raman spectra of 2D substrates obtained before and after GaN NW growth in line with the findings reported in extant literature[161-167] and Raman spectra of bare MoS$_2$ and WSe$_2$ as depicted in Figure 4.15, demonstrating that initial 2D substrate crystallinity was well maintained, and GaN growth temperature does not affect the nature of the 2D materials. However, a slight shift was observed, which can be associated with the change in 2D bond length, owing to the heterostructure formation.[138]

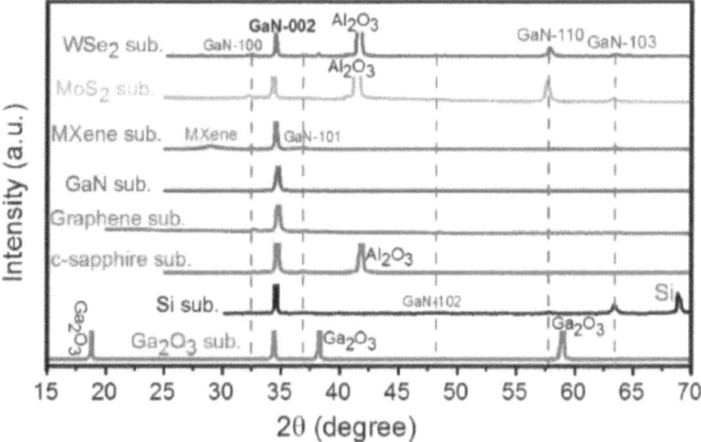

Figure 4.13. XRD patterns produced by GaN NWs grown on all substrates.

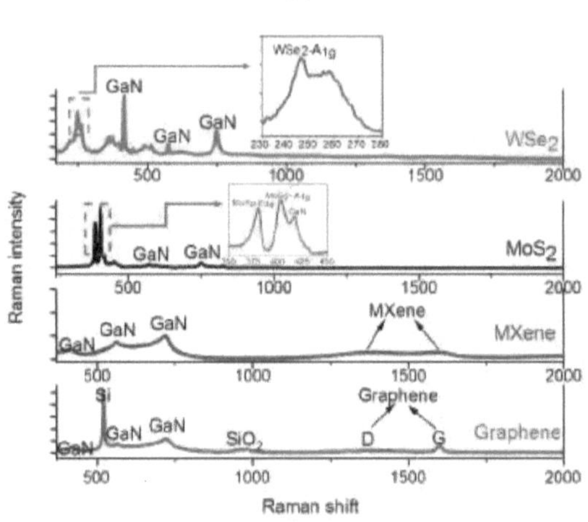

Figure 4.14. Raman spectra of NWs grown on all 2D substrates.

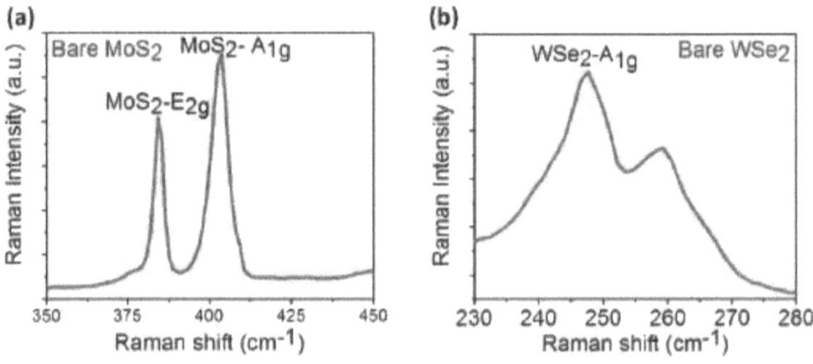

Figure 4.15. Raman measurements of bare TMD materials (a) MoS_2 and (b) WSe_2.

ii)   *Optical properties*

To study the optical properties, we performed RT absorption and RT photoluminescence micro-PL (μ-PL) measurements on the GaN NWs. The findings reported in Figure 4.16 confirm that GaN NWs grown on Si and sapphire produce a sharp absorption edge. Typical absorption curves were obtained for NWs grown on all substrates. The absorption results confirm that the bandgap energy is around ~3.58 eV and ~3.68 eV for samples grown Si and c-sapphire substrates, respectively. The observed slight shift in the absorption edge for NWs grown on different substrates can be due to the difference in the NW diameter (i.e. different compressed strain) grown on different substrates,[149, 168] as confirmed by XRD and SEM results.

On the other hand, the RT μ-PL spectra shown in Figure 4.17 exhibit sharp dominant GaN near band edge (NBE) emission at 346–350 nm (3.58–3.54 eV) for all NWs grown on all substrates, along with a weak defect band (yellow band), confirming superior optical quality, which is in line with the XRD, the sharp absorption edge and STEM results (that will show later in chapter 5), and demonstrating the success of our growth method. The PL spectra of optimized samples show strong NBE emission with a weak defect band, confirm high crystal quality.[37, 41] Although there is a negligible defect yellow band, NBE emission is broad, which can be due to high oxygen impurities[41] as oxygen impurities were detected in the GaN target (as will be shown in Chapter 5 section 5.3.1). Therefore, the broadening is not due to low crystal quality.

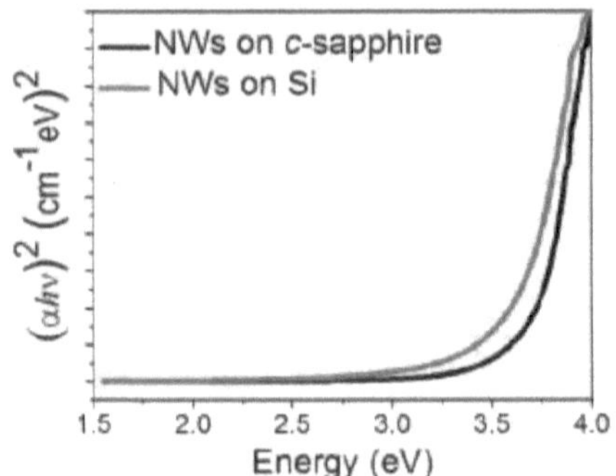

Figure 4.16. Tauc plot of the GaN NW energy bandgap when c-sapphire and Si are used as the substrates.

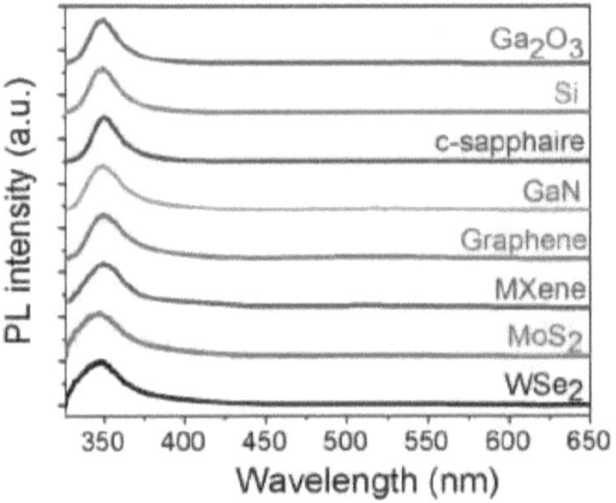

Figure 4.17. PL spectra of GaN NWs grown on different substrates.

4.3.5. Large scale growth (4-inch wafer)

To examine the possibility of obtaining high-quality GaN NWs across large wafer substrates using our growth method, as this procedure is essential for large-scale industrial production, we also grew high-quality vertically-aligned GaN NWs on a 4" Si substrate, as shown in Figure 4.14, indicating that a slight change is observed in the NW shape without compromising on the quality. The room temperature (RT) micro-photoluminescence (μ-PL) spectra taken from different locations across the wafer show a dominant GaN near bandgap edge (NBE) peak with a weaker yellow band, as illustrated in Figure4.15, confirming the high quality of NWs across the entire wafer surface,[169] Thus, our method is suitable for large-scale applications, as many high-quality GaN NWs grown on 4" substrates can be formed from one target.

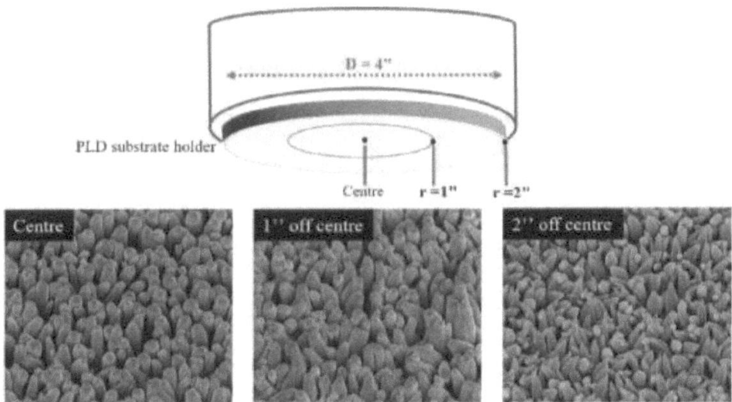

Figure 4.18. SEM images were acquired at different positions across a 4" substrate.

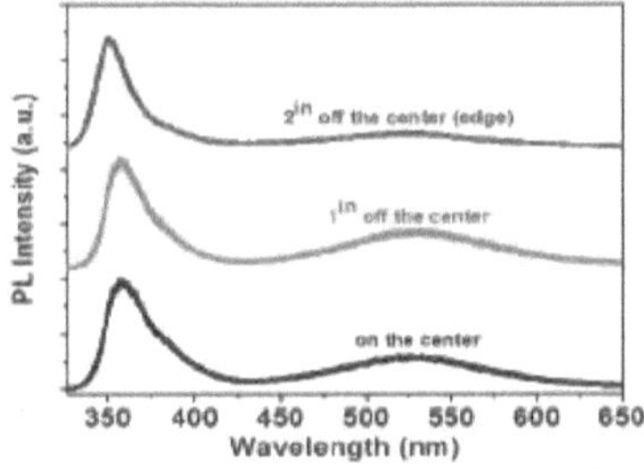

Figure 4.19. PL spectra of GaN NWs growth at different positions across a 4" substrate.

4.4. Summary

In summary, for the first time, we have developed a novel one-step method for growing high-efficiency vertical single-crystal self-assembly GaN NWs without a catalyst on different common bulk and emerging 2D substrates by using PLD. Thus, eliminating the need for costly or complex fabrication processing. The optimal conditions to grow GaN NWs were 150–250 mTorr nitrogen pressure (P_{N2}); 1 ± 0.05 J·cm^{-2} beam fluence focused on the GaN target surface (with 40,000–100,000 laser pulses delivered at a frequency of 10 Hz); ~9 cm vertical distance between the substrate and the target; and 850 °C growth temperature. GaN NWs length varied from 1.5 μm to 3.3 μm based on the number of laser pulses and their high quality is confirmed by several measurements such as PL, XRD, Raman shift, and RT absorption measurements.

Chapter 5

Growth Mechanism of high-optical efficiency dislocation-free GaN NWs without catalyst or seeding

5.1. Introduction

It is known that polycrystalline formation on a substrate disables the dislocation growth in the interface between grown materials and substrates.[170, 171] However, the strategy of growing high optical and structural quality single crystal GaN on polycrystalline GaN has not been demonstrated yet. Overcoming the issues of lattice mismatch and the effect of TD on the interface between the substrate and GaN will allow GaN to grow on any substrate, including emerging 2D substrates for flexible applications.

This chapter shows the growth mechanism of state-of-the-art NWs in several substages. We have resolved these GaN challenges (that were explained in Chapter 2 Section 2.8) by developing a methodology that does not depend on the lattice mismatch. In addition, we have overcome this catalyst layer issue by demonstrating that any bulk or 2D substrate can be used for growth irrespective of the intended application by using the PLD technique. Advanced characterizations confirm the superior optical and structural quality of these single-crystal NWs, demonstrating that this growth strategy can be generalized for a wide range of semiconductor applications, including flexible devices.

5.2.   Experimental method

5.2.1. The sample preparation

The GaN NWs growth conditions and substrates preparation have been described in Chapter 4, Sections 4.3.1 and 4.3.2.

5.2.3. Structural Characterizations

Titan Themis Z (40-300) Thermo Fisher scanning transmission electron microscopy (STEM) and high-resolution STEM (HR-STEM) was used. Energy-dispersive X-ray (EDX) spectroscopy was performed in the TEM apparatus to determine the NW composition. Also, fast Fourier transform (FFT) and selected area electron diffraction (SAED) in the TEM to clarify the polycrystalline and single-crystal structure of our NWs. TEM lamellae were prepared by focused ion beam FEI Quanta 3D (FIB-SEM). Depth profiling experiments were performed using a Dynamic Secondary Ion Mass Spectroscopy (SIMS) instrument (from Hiden analytical company, Warrington-UK) operated under ultra-high vacuum conditions, typically 10^{-9} torr. A continuous Ar^+ beam of 4 keV energy was employed to sputter the surface while the selected ions were sequentially collected using a MAXIM spectrometer equipped with a quadrupole analyzer. The raster of the sputtered area is estimated to be $750 \times 750\ \mu m^2$. In order to avoid the edge effect during depth profiling experiments, it is necessary to acquire data from a small area located in the middle of the eroded region. Using an adequate electronic gating, the acquisition area from which the depth profiling data are obtained was approximately $75 \times 75\ \mu m^2$. Assuming a constant sputtering rate, the conversion of the sputtering time scale to the depth scale was carried out by measuring the depth of the crater generated at the end of the depth profiling experiment using a stylus profiler (from Veeco company). SIMS measurements carried it out in the KAUST Core Lab by Dr. Nimer Wehbe.

5.2.3. Optical characterizations

Power–dependent photoluminescence (PDPL) measurements at 10 K and room temperature (290 K) were carried out using a 266 nm laser (Teem photonics SNU-20F-10x), whereby the obtained PL emission signal was detected by a highly sensitive QE Pro spectrometer with the acquisition time fixed to 500 ms for 30 integrations. We used a closed-cycle cryostat for low-temperature PL measurements. RT cathodoluminescence (CL) hyperspectral imaging was carried out using a custom-built acquisition system in an SEM, using 5 kV and an acquisition time of 100 ms/pixel. Temperature-dependent PL measurements at 5 K and RT was carried out by using a 325 nm continuous-wave (CW) He–Cd laser, while an Andor spectrograph connected to a charge-coupled device camera was used to collect the data.

5.3. Results and discussions

5.3.1. Advanced High-resolution STEM Analyses

i) *Polycrystalline wetting layer*

We carried out extensive STEM, HR-STEM, and EDX measurements on the GaN NW grown on all substrates to understand the growth mechanism. Figure 5.1a shows the STEM image of vertical self-assembly GaN NWs grown on all substrates. As shown in the STEM images in Figure 5.1b, 5.2a, and 5.2b, polycrystalline GaN WL was formed below GaN NWs grown on the Si (100), sapphire, and GaN substrates, respectively. Similar results were obtained for other substrates, as shown in Figure 5.3a and c, as well as Figure 5.4a.

FFT pattern confirms this finding (polycrystalline structure) as shown in the inset in Figure 5.1b. The HR-STEM images reveal that polycrystalline WL consists of several single-crystal grains (coherent crystallite) with different orientations, which is in line with the different minor GaN peaks observed in the XRD scan shown in Figure 4.9 in Chapter 4 Section 4.3.4. STEM images of a wider area indicate that the WL thickness ranges from 200 to 500 nm, as shown in Figure 5.5a–c, as well as in Figure 5.6a.

Interestingly, the HR-STEM images of the interface between the polycrystalline GaN WL and the substrates reveal that, during one-step PLD growth, an *in-situ* homogeneous nano-layer of uniform ~ 2 nm thickness was also formed. This nano-layer acts as a catalysis layer and assists in the WL formation and NW growth on all substrates. The uniformity of the interface nano-layer assists in obtaining homogenous NWs. To further confirm the nature of the interface nano-layer, the EDX maps of the chemical constituents of the sample based on Si substrate were performed, as shown in Figure 5.1c. This nano-layer contains Ga, N, and silicon oxide (e.g. SiO_2), as demonstrated by the atomic percentage profiles presented in Figure 5.1d. A similar nano-layer with oxide in its composition has been formed during the growth on the interface between the WL and sapphire and other bulk substrates, as shown in Figure 5.2a and b, respectively. This oxygen-rich *in-situ* nano-layer is a result of the oxygen dopants in the PLD GaN target. SIMS measurements showed a strong oxygen signal from the PLD GaN target as shown in figure 5.1e (Note that SIMS measurements were carried out for all NW samples, confirming the presence of oxygen dopants in all samples).[172] This results indicates that, during the growth process, the oxygen atoms move to the interface, as it is known that poorly soluble impurities prefer to accumulate either on the material surface or on the interface between the materials.[148, 149] Similar results have been previously observed for Gd-doped ZnO NWs and NTs.[148, 149] In these previous studies, as Gd-doped ZnO target was used, Gd dopants accumulated above the substrate and formed Gd *in-situ* layer in the interface between the substrate and the ZnO WL, assisting in forming vertical ZnO NWs/NTs during PLD growth.[148, 149]

As GaN NWs grown on MXene and TMD (e.g., MoS_2) are emerging substrates, they were also used for growing GaN NWs to demonstrate the wide-ranging applicability of our method. Figure 5.3a–d show the cross-sectional HR-STEM images and related EDX maps, as well as the related elemental profiles of samples grown on MXene and TMD (e.g. MoS_2), respectively, while Figure 5.4a and b shows NWs grown on graphene. HR-STEM analyses reveal the formation of a polycrystalline WL above all 2D substrates. In this case, the 2D substrates (graphene, MXene, MoS_2, and WSe_2) may serve as an *in-situ* nano-layer in bulk substrate. For samples grown on 2D substrates, the semiconductor NW formation has been attributed to the weak quasi Van der Waals interactions between 3D material and 2D substrates.[173, 174]

ii)   *Absence of TD defects at the NW–substrate interface*

Surprisingly, the HR-STEM images of the GaN polycrystalline WL on different substrates demonstrate a complete absence of TD between the GaN and all substrates, as shown in Figure 5.1b, 5.2a and b, 5.3a, 5.3c, and 5.4a, irrespective of the degree of lattice match/mismatch between GaN WL and the chosen substrate. STEM images are shown in Figure 5.5a–c as well as in Figure 5.6a, pertain to a larger WL area on different substrates, further confirming that TDs have been eliminated. Note that the large vertical lines in Figure 5.6a are due to the FIB ion beam passing through in-between the NWs patenting toward the substrate, as shown in Figure 5.1a as well. Such a complete absence of TDs in polycrystalline materials has been attributed to the nature of WL polycrystallinity, which prevents the lattice mismatch effect from occurring during growth. It is known that polycrystalline layer formation on a substrate disables the dislocation growth in the interface between grown materials and substrates due to the different orientation of the grains, resulting in no crystallographic relation between the polycrystalline layer and the substrate, in contrast to the epitaxial growth of crystalline layers.[145, 146] This is a significant achievement for GaN and this field, given that TDs are always present in the interface between these substrates and GaN and other III-nitrides.[127, 128]

We studied the grain type and size statistically on the samples grown on different substrates as shown in Figure 5.6d and 5.6e. Our data suggested that statistically no significant difference was observed between grain size for samples grown in different substrates. In both samples, going from the substrate to the top of the NWs, we observe the grain size growing larger eventually turning into a single crystal until the end of NW.

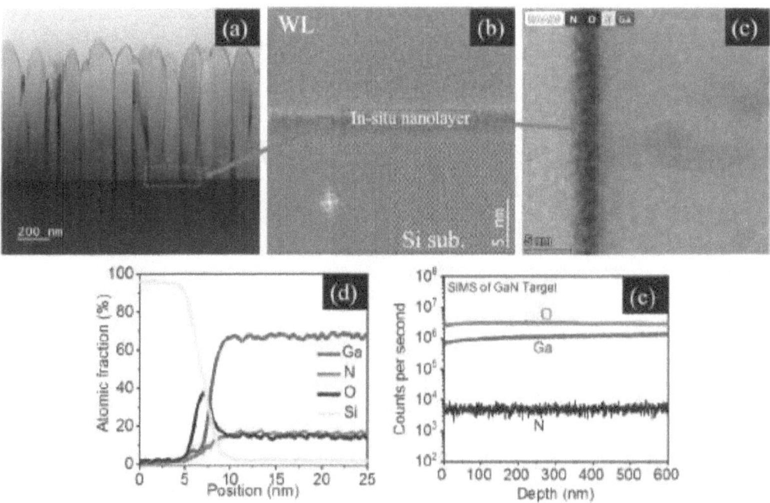

Figure 5.1. (a) The STEM image of the vertical self-assembly GaN NWs has grown on the Si substrate. (The vertical lines in the substrate arise due to the ion beam penetrating the substrate during FIB lamella preparations). (b) HR-STEM images of polycrystalline GaN WL, and *in-situ* interface layer formed between WL and Si substrate, (the inset shows FFT patterns of the polycrystalline WL). EDX (c) map and (d) profile, revealing *in-situ* nano-layer composition on Si substrate. (e) SIMS measurement of GaN target showing sufficient oxygen incorporation. (All NW samples showed oxygen incorporation as well).

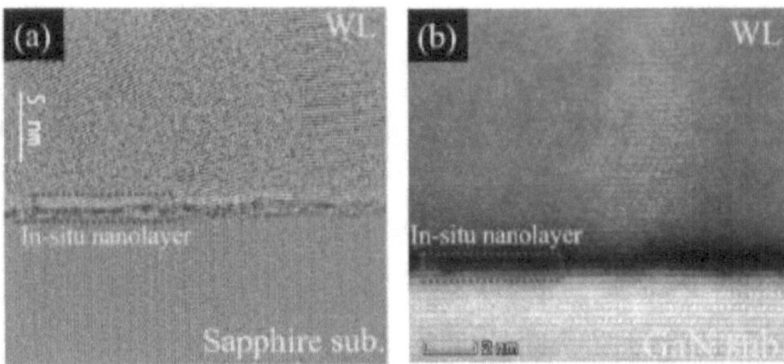

Figure 5.2. HR-STEM images of polycrystalline GaN WL, indicating a complete absence of dislocations, and in-situ interface layer formed between WL and (a) c-sapphire and (b) GaN substrates.

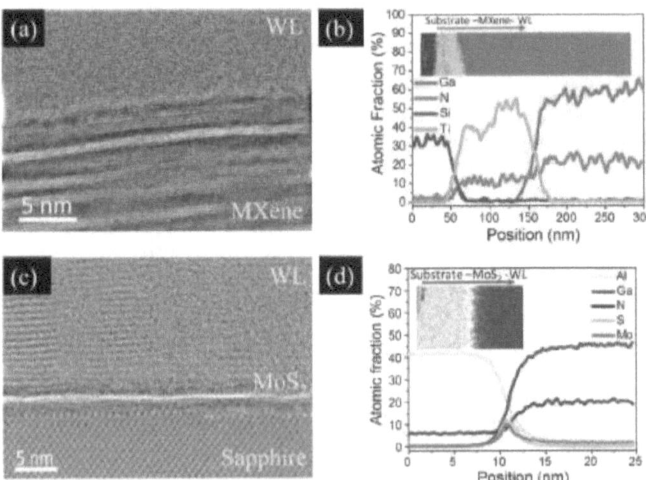

Figure 5.3. (a) HR-STEM images of the interface between GaN WL and MXene substrate. (b) The EDX map and profile of the STEM image of MXene. (c) HR-STEM image of the interface between GaN WL and MoS_2 substrate. (d) The EDX map and profile of the STEM image of MoS_2.

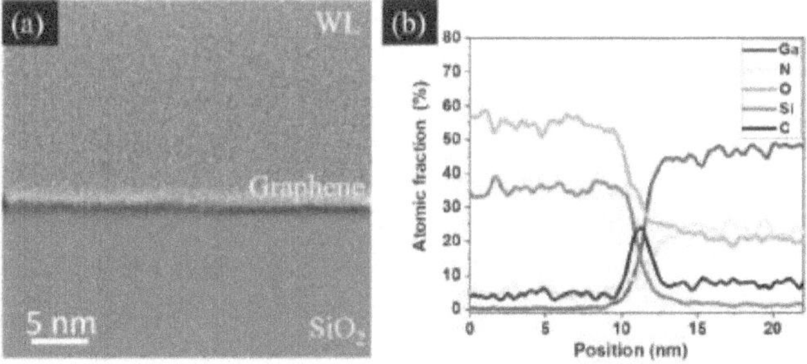

Figure 5.4. (a) HR-STEM images of the interface between GaN WL and graphene substrate and (b) EDX profile of interface composition on the graphene substrate.

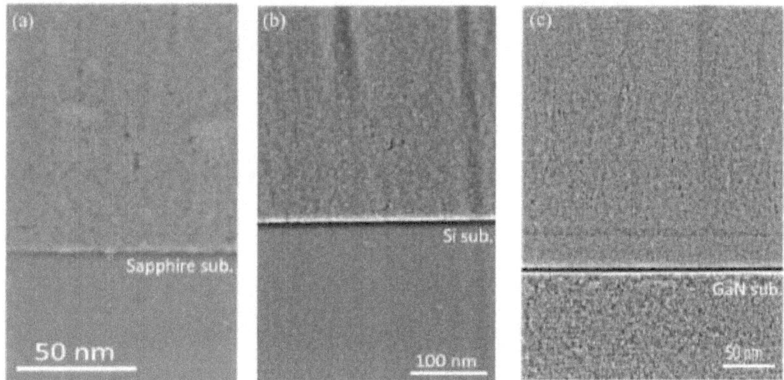

Figure 5.5. STEM images of amorphous GaN WL on (a) sapphire, (b) Si, and (c) GaN substrates. The vertical lines are the effect of the ion-beam used by FIB during lamella preparation.

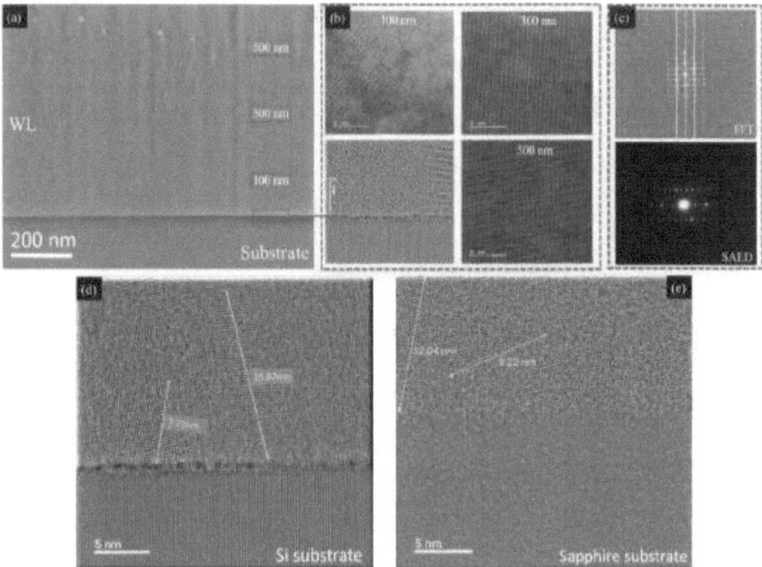

Figure 5.6. (a) STEM image of a full WL layer. (b) The corresponding HR-STEM images taken at the interface (the bottom-left panel), and at 100 nm (top-left), 300 nm (top-right), and 500 nm (bottom-right) height above the interface, confirming that the polycrystalline nature diminishes as the distance away from the substrate increases, retaining single-crystalline structure in the NWs. (c) The corresponding SAED and FFT patterns are taken from the NW bottom, confirming its single-crystal form. The grain size in the GaN WL for samples grown on (d) Si and (e) sapphire substrates.

iii)   *Single-crystalline nature of GaN NWs*

HR-STEM analysis demonstrated the high quality and single-crystal structure of vertically aligned wurtzite GaN NWs for all samples grown on all bulk and 2D substrates, as shown in Figure 5.7a–d and 5.8a–d. The single-crystalline structure was further confirmed by

SAED and FFT patterns, as shown in Figure 5.6c. (We show here Si and sapphire samples as examples of bulk substrates, and MXene and MoS_2 as examples of 2D substrates).

To explore the process underpinning the transformation from polycrystallinity (in WL) to single-crystallinity (in NWs), we carried out STEM measurements at different locations situated progressively vertically further away from the substrate/WL interface, as shown in the area presented in Figure 5.6a. The HR-STEM images provided in Figure 5.6c (surrounded by the red dotted line) show the crystalline nature of the corresponding areas. These HR-STEM images reveal that the polycrystalline characteristic diminishes as the distance from the substrate increases, moving towards the GaN NWs along the c-axis (see Figure 5.6a–c), whereas the single-crystalline structure along c-direction is retained in the NWs (as shown in Figure 5.7a–4d and Figure 5.8a–d). Such behavior is due to the via Stranski–Krastanov (SK) growth mode, as will be discussed below.

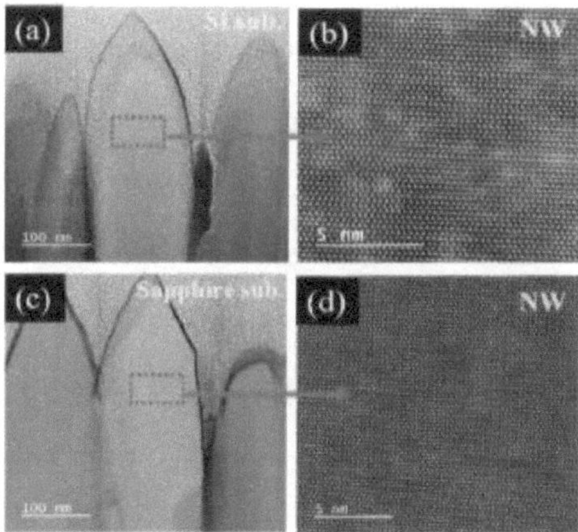

Figure 5.7. (a) STEM image for GaN NWs on Si substrate. (b) HR-STEM image showing the crystalline structure of GaN NWs on Si. (c) STEM image for GaN NWs on the c-sapphire substrate. (d) HR-STEM image of the crystalline structure of these GaN NWs.

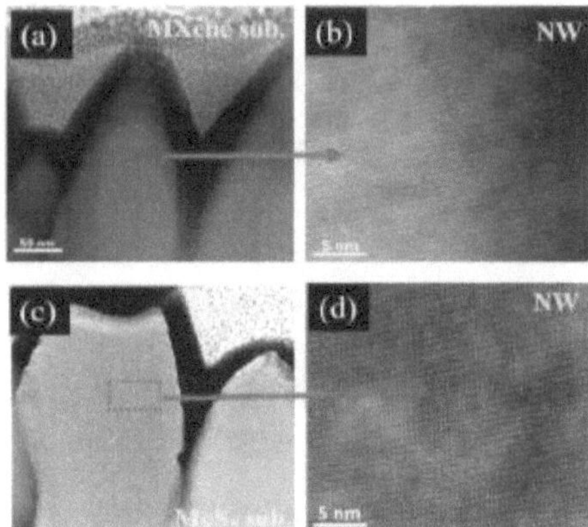

Figure 5.8. (a) STEM image of GaN NWs grown on the MXene substrate. (b) HR-STEM image of the single-crystal structure of NWs on MXene. (c) STEM image of GaN NWs grown on MoS_2. (d) HR-STEM image of the single-crystal structure of NWs on MoS_2.

5.3.2. Growth mechanism of single-crystal NWs via *in-situ* polycrystalline WL using the surface energy concept

The surface energy is associated with the intermolecular forces at the interface between two surfaces.[71, 149, 175] Different processes (growing nucleation) governed by the substrate surface energy and the interface energy between the substrate and the growing islands are employed to control the nucleation growth mode, i.e., either Volmer-Weber (VW), resulting in islands or 3D structures; Frank-van der Merwe (FM), resulting in a layer or a 2D structure; or SK, leading to layer-plus-island growth mode, as explained in Chapter 2 Section 2.6.1.[71] Thus, the high-quality vertically-aligned GaN NWs on different 2D and bulk substrates without catalyst is governed by the surface energy of the grown materials, which is modified by controlling the energy (E) of the charged species landing on the substrate, resulting in SK, VW, or FM growth mode. Therefore, the following E equation was previously used to explain how the species energy governs the formation of 3D (island) and 2D (film) structures.[72, 149] The species energy can be defined using the following equation, (see Chapter 2 Section 2.6.2 for deriving the equation):[72, 149]

$$E = E_o \, exp \left(-\frac{\pi a_0^2 P d \sqrt{2}}{kT}\right); \qquad\qquad 2.19$$

where E_o represents the initial energy of the charged species in the plume emerging from the target surface, d is the distance between the target and the substrate, the growth temperature (T), and nitrogen pressure (P). Also, a_o denotes gas molecule diameter, and k is Boltzmann constant.

In the present study, 850 °C growth temperature was adopted, as this is the maximum that can be used in our PLD system, and a similar temperature has been used in other works using for MBE technique. However, using lower temperature was not suitable for NW growth, as shown in Figure 4.2 in Chapter 4 Section 4.3.3. Based on Equation (2.19), under the given PLD conditions, which corresponds to $P_{N2} \geq 150$ mTorr for $d = 8.5\text{--}9$ cm and T = 850 °C, the optimized initial species energy E_o was equivalent to laser fluence of 0.95–1.05 J/cm^2. Under these T and d conditions, relatively high P_{N2} (≥ 150 mTorr) and/or relatively low E_o of the ablated species can result in high scattering due to a large number of collisions with nitrogen molecules, further decreasing E of the charged species arriving at the substrate surface,[72, 149] thus allowing NW formation on WL without catalysis or seeding by satisfying the SK nucleation condition. On the other hand, using the given T and d values, we found that when E exceeds the optimized value—which is achieved by reducing P_{N2} or by increasing laser fluence on the target, according to Equation (2.19) a film structure is formed, as the ablated species in the emitted plume arrive at the substrate with very high kinetic energy E, resulting in a continuous film (satisfying the surface energy conditions for the FM growth mode, as described in Figure 4.3e in Chapter 4 Section 4.3.3).[72, 149] Figures 4.3e and 4.4b show the granular film is produced using higher energy E and/or lower pressure P_{N2} that of optimized values, respectively. Furthermore, we found that the number of pulses should exceed 40000 pulses to achieve high NW quality, as shown in Figure 4.5 in Chapter 4 Section 4.3.3. These findings indicate that all aforementioned parameters need to be carefully adjusted dependently to obtain optimized conditions.

These findings further revealed that the growing nucleation mode needs to be controlled by modulating the stress (which can be achieved by modifying the landed species energy E), as well as the formation of the oxidation nano-layer (or 2D substrate) materials between the substrate and the WL to satisfy the SK nucleation ($\gamma s > \gamma i + \gamma n$, where γs is the surface energy of the substrate, γn pertains to the growing nucleation, and γi represents the interface energy between the substrate and the growing islands (see Chapter 2 Section 2.6.1)).[71, 149] In the SK mode, the 3D islands are of great practical importance for the growth of self-assembly nanostructures such as QDs or NWs.[71, 149, 176] In SK growth, the strained layer (WL) on the substrate is grown first, then the growth switches from layer-by-layer to 3D island after strain relaxation. In this case, the islands continuously relax owing to lattice distortion in the growth direction, due to which the surface energy changes from the FM mode (layer) to SK mode (layer (WL) + island). At this point, this 3D island should be dislocation free (known as coherent islands).[170] In other words, changes in strain can also lead to a growth mode transformation,[170], and entropy is found to play a crucial role in stabilizing the WL in the SK mode.[176] In our growth strategy, the polycrystalline layer acts as the WL that assist in generating 3D island, facilitating NWs formation.[145] In polycrystalline materials, grain boundaries have been proposed as a possible source of intrinsic stress. According to Hoffman,[146] the interatomic forces at the boundaries tend to close any existing gap, due to which the neighboring crystallites (e.g. single-crystal grains) are strained.[145] Consequently, the polycrystalline layer does not exhibit dislocations at either of its interfaces, which in line with our resulted polycrystalline WL, allowing the 3D island characteristics and leading to SK growth mode conditions.[170] Thus, these single-crystalline grains may play a role in the absence of TD at the WL-substrate interface,[146, 170] while initiating single-crystal NW formation.[177]

Our GaN NW growth strategy using PLD can be applied in other techniques, such as sputtering or MBE. This can be achieved by controlling the landed species energy (E) using the same theory explained in this section.

5.3.3. High optical efficiency

i)   *Emission properties*

To elucidate the optical quality, we studied temperature- and power-dependent PL measurements on the GaN NWs. To elucidate the reason behind the broadening of the intense GaN emission with a negligible yellow band, RT cathodoluminescence (CL) hyperspectral imaging was performed to spatially resolve the emissions from the polycrystalline WL and NW regions. CL maps were acquired from the cross-sectional SEM image GaN NW sample, as shown in Figure 5.9a. Figure 5.10 shows the averaged CL map over the full image presented in the inset, taken from the top to the bottom of the sample. The spectrum provided in Figure 5.10 reveals two peaks centered at ~344 nm (~3.6 eV) and ~326 nm (~3.8 eV). The averaged CL map indicates that the 3.6 eV peak is emitted from NWs as shown in Figure 5.9b, which is in the same position in the PL spectra, indicating that the dominant emission is generated from NW, whereas the ~3.8 eV peak is emitted from the polycrystalline WL (PL spectrum shown in Figure 5.9c). Thus, the broadening of NBE emission can be due to the overlap of these two peaks.

The NBE peaks originating from WL are blue-shifted and broad (as shown in the PL spectrum illustrated in Figure 5.9c) which can be a result of the single-crystal nano-grains. Widmann et al.[178] found that the PL spectra of self-assembled GaN quantum dots contain a similar broad NBE intense emission centered at 3.8 eV, and ascribed the broadening to significant QD size fluctuations. Thus, the single-crystal nano-grains in the polycrystalline WL may exhibit quantum dot-like behavior, as the average grain size remains below 10 nm (i.e., it does not exceed the exciton diameter), as shown in the STEM images of the WL, leading to exciton confinement. Therefore, as excitons are confined and localized in these nano-grains, NBE blueshift occurs,[178] while the absence of a defect band (Figure 5.9c) confirms the high quality of the WL layer.

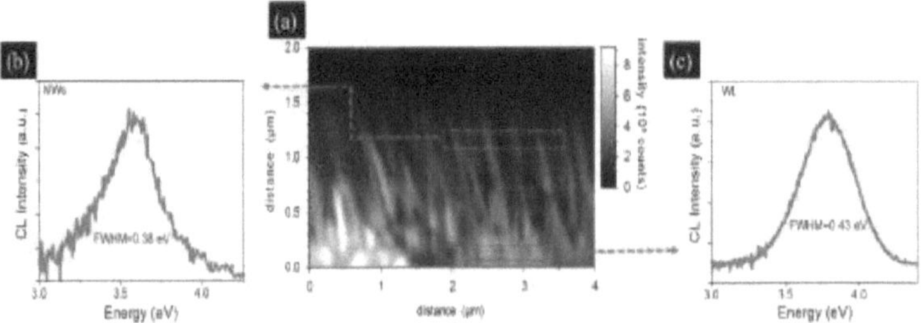

Figure 5.9. (a) CL map of the side wall of an NW sample, and the CL spectra pertaining to (b) NW and (c) WL.

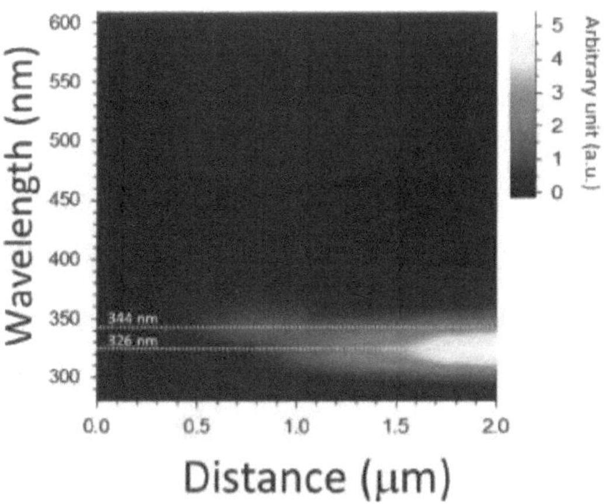

Figure 5.10. Averaged RT CL spectral map (over full image width (wavelength)) vs distance from the top to the bottom of the NW sample, including the WL area.

ii) *Internal efficiency analysis and dominant radiative recombination*

The intense emission produced by both WL and NWs can lead to high optical efficiency and dominant radiative recombination. Thus, temperature–and power–dependent PL measurements were carried out to elucidate these findings. Figure 5.11a shows that the NBE emission is intense at both RT and 5 K. The tail shown at the lower energy side in Figure 5.11a was attributed to the presence of stacking faults in GaN NW, which is a typical NW signature.[179] The IQE was estimated by evaluating the ratio of the integrated intensity at 5 K to that at RT, assuming that all nonradiative centers are frozen at 5 K.[180] The IQE was found to be 43% for the sample grown on Si (Figure 5.11a), while it ranged from ~ 40% to 45% for samples grown on different substrates, demonstrating high UV IQE efficiency and superior quality.

As thermal IQE is not very accurate and is affected by laser power, to further confirm the high efficiency of GaN NWs, we calculated IQE for the sample grown on Si (as it has the greatest lattice mismatch with GaN compared to other substrates) using Shockley−Read−Hall (SRH) method through power-dependent photoluminescence (PDPL) measurements, which is considered the most accurate and reliable method for calculating IQE at RT.[181] The PDPL spectra shown in Figure 5.11b, which were obtained under different injected laser power, indicate that the NBE emission consists of two overlapping peaks, which is in agreement with the CL measurements. The IQE was calculated by applying the ABC model to the PDPL data, as described in Section 2.3.5 in Chapter 2:[56, 181]

$$G_{opt} = An + Bn^2 + Cn^3; \qquad\qquad 2.6$$

where An represents the Shockley−Read−Hall (SRH) nonradiative recombination rate, Bn^2 denotes the radiative recombination rate, and Cn^3 is the carrier Auger-like nonradiative recombination rate. This equation shows that the three regimes can be represented via the relationship between the injected current density (G) and carrier concentration (I), whereby $I_{PL} \alpha G^k$ (Chapter 2 Section 2.3.6, Equation 2.12). For the nonradiative SRH recombination regime, at low carrier injection, $k = 2$ would apply. Similarly, at high carrier concentration $I_{PL} \alpha G^{k=2/3}$ for Auger recombination regime, whereas $k = 1$ holds when total radiative recombination regime is applied.[54]

Consequently, based on the analysis shown in Section 2.3.6 in Chapter 2, the IQE at steady state can be expressed as:

$$\eta_{IQE} = \frac{Bn^2}{G_{opt}} = \frac{Q_2 I_{PL}}{G_{opt}}; \qquad\qquad 2.13$$

where Q_2 is a constant (Chapter 2, Section 2.3.6). The power-dependent IQE as a function of power density (the injected carrier density) at RT is shown in Figure 5.12a. The IQE curve depicted in Figure 5.12a indicates that the IQE increases rapidly with the excitation energy density due to the saturation of the nonradiative recombination centers by generated carriers,[182] whereby its maximum value (65%) is achieved at $\sim$70 kW/cm^3, demonstrating significant predominance of radiative recombination. At higher excitation energy densities, the IQE starts to decline, reaching $\sim$47% at 1.2×10^3 kW/cm^3, which is still considered high for GaN UV emission,[54, 182], and is in line with the thermal IQE. The concave IQE droop behavior of these NWs further confirms the dominant radiative recombination process even after the droop,[43] indicating that these NWs can be potential candidates for the development of high-efficiency LEDs.

To determine the radiative recombination contribution and elucidate the reasons behind the droop, we plotted log (I_{PL}) as a function of log (G_{opt}) (power density) at RT and 10 K. According to Equation 2.12 provided in Chapter 2 Section 2.3.6, the dominance of radiative recombination is demonstrated when $I \propto G$ ($I \propto n^2$), resulting in a rapid increase in the IQE,[54] as the majority of the contribution comes from localized excitons.[148] At RT, Figure 5.12b shows that $k \sim 1.2$ is obtained under a low excitation power density, which further confirms the significance of the radiative recombination contribution in the total recombination process, with some defect-related nonradiative recombination. On the other hand, at the higher values of the IQE range, as can be seen in Figure 5.12b, at $\geq$ 30 kW/cm^3 before the droop regime, $k = 1$, confirming a complete radiative recombination process due to saturation of nonradiative centers at high excitation power density. However, in the efficiency droop range, the slope is $\sim$ 0.89, indicating that dominant radiative recombination is accompanied by minor Auger recombination contribution, and thus revealing that Auger recombination can be the main reason behind the slight IQE droop. At low temperature (10 K), the k value of 1−1.1 confirms that radiative recombination completely predominates in both low and high excitation regimes.[22, 54, 182] These results demonstrate that these highly efficient NWs have strong potential for use in a wide range of applications.

Considered jointly, the findings reported here confirm that our growth method eliminates TDs in NWs grown on a wide range of bulk and 2D substrates of any type (conductive, insulator, transparent, or non-transparent), allowing it to be adopted in the development of high-performance GaN-based devices intended for a wide range of III-nitride applications. Thus, our novel growth strategy can be applied for growing high-quality NWs directly on any substrate and can pave the way for the use of III-nitrides for several potential applications. We believe that many other substrates–fixable as well as polycrystalline or amorphous such as silica (SiO_2)– can be a very good substrate for our GaN growth strategy.

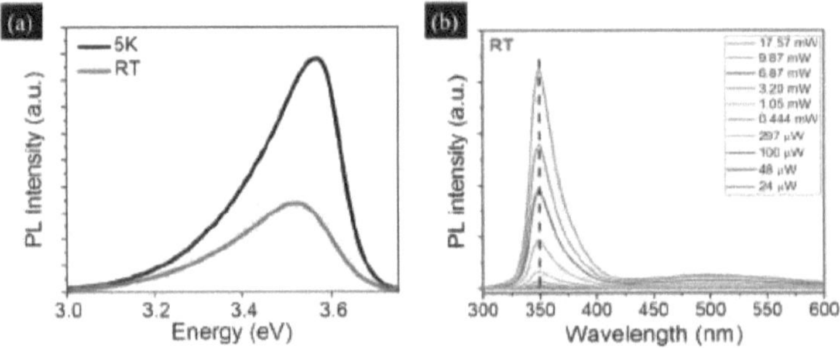

Figure 5.11. (a) Temperature-dependent PL spectra of GaN NWs at RT and 5 K. (b) Power-dependent PL spectra of GaN NWs at 10 K and RT.

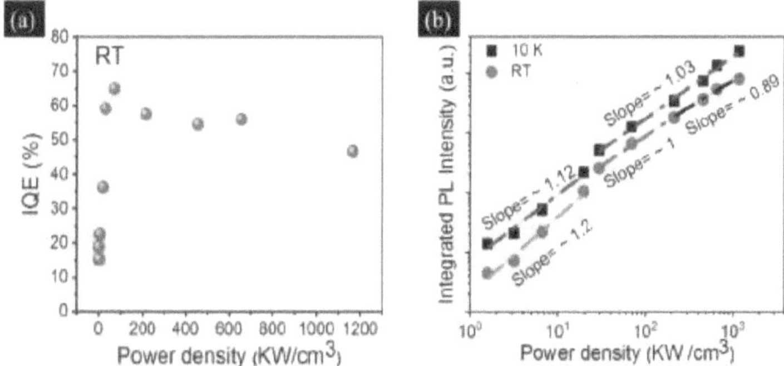

Figure 5.12. (a) IQE calculated from the PDPL integrated intensity at RT as a function of excitation power density, and (b) (log-log) integrated RT PL intensity output as a function of power density.

5.4. Summary

In summary, we have resolved the main GaN growth challenges by developing a methodology that does not depend on the lattice mismatch, as no dislocation defects and no catalyst layer. HR-STEM demonstrated single-crystal state-of-the-art GaN NWs. STEM images showed an *in-situ* layer between bulk substrates and GaN WL, that worked as a catalyst layer. For the case of the 2D substrate, the 2D layer made the same action as the *in-situ* layer for bulk substrates. STEM findings indicate the complete absence of dislocations between GaN WL and all substrates. As well as high quality of TD-free single-crystal GaN NWs grown on different substrates as showed in STEM images. The absence of dislocations in the high optical and structural quality NWs as well as nano-grains in WL further enhanced efficiency. Carrier dynamics measurements on all samples confirmed superior UV efficiency, while a slight droop at high excitation densities is ascribed to a minor contribution from Auger recombination at high carrier injection rates. These findings open up new horizons in research on GaN-based emitting devices, including UV vertical emitting laser diode and flexible devices.

Chapter 6

Light-emitting devices structure based on PLD GaN NWs

6.1. Introduction

The market demands low-cost LED devices based on III-nitride multiple quantum wells (MQWs), as such structures are suitable for large-scale applications. However, the cost of the substrate material and required processing must be reduced to obtain inexpensive light emitters for large scale product.[183] Usually, the commercial blue and green GaN-based light emitters are solely grown on sapphire and SiC substrates.[183] InGaN/GaN MQWs are commonly employed as the active layer in nitride-based LEDs and laser diodes (LDs) because of their high radiative recombination efficiency and their capability of emitting in a wide spectral range, spanning almost the entire visible spectrum and extending into the ultraviolet region.[184]

Several optoelectronic devices and device structures, including LEDs, are based on InGaN/GaN multiple MQWs.[185] Typically InGaN/GaN MQWs are grown on a thick GaN template layer deposited on the sapphire substrate.[184] InGaN LEDs have also been grown on Si at low temperatures using a thick AlN buffer layer and an AlGaN/GaN strained-layer superlattice (SLS).[186] Besides, such LEDs suffer from both high operating voltage and high series resistance, which result from the insulating AlN layer and the large band offset at the AlN/Si interface.[186] These issues are presently difficult to overcome, as the growth of high-quality InGaN/GaN MQWs emitting light in the green-to-red spectral range is still a challenge,[184] due to which the resulting devices are characterized by low efficiency and efficiency droop. These phenomena are attributed to the presence of polarization fields, Auger recombination, poor hole transport, defects/dislocations, and/or electron leakage and overflow.[187] It is known that cost-effective GaN grown on Si is of very poor quality due to the significant lattice mismatch between GaN and Si, resulting in high dislocation density, as noted previously.[127] Thus, the dislocations penetrating the quantum wells (the active layer) reduce the device efficiency significantly.

Since 2004, several III-N NW-based LEDs that integrate a great number of p–n nanodiodes connected in parallel have been developed,[188, 189] including IIII-nitride MQWs-based LEDs.[188, 189] In NW structures, the dislocation density is reduced,[190, 191] while a reduced strain distribution in the nanostructures also leads to a weaker piezoelectric polarization field. The reduced strain in NWs also allows for incorporating a higher InN fraction into InGaN/GaN MQWs, which is an advantage for green and red emitting devices.[192] Other advantages of GaN NWs include high light extraction efficiency and compatibility with low-cost, large-area silicon substrates.[191]

Vertically aligned p-GaN/In_xGa_{1-x}N/GaN MQW/n-GaN NW-based LED structures grown on Si substrates have been obtained by MOCVD and MBE techniques, using both top-to-bottom and bottom-to-top approaches.[183, 191, 193] The spectral range of LED emission was changed from the violet to the green region in InGaN LEDs by controlling the indium composition.[194, 195] For example, In composition was nearly zero (~0.05) for UV LEDs, while it was 0.18–0.2 for blue LEDs and 0.25–0.45 for green LEDs.[194-196] In extant studies, InGaN/GaN MQWs have been grown on GaN nanowires/nanorods for application in LEDs and lasers.[197] Similarly, Jung et al. investigated the possibility of In_xGa_{1-x}N/GaN MQWs nanowire on GaN NW array.[198] These authors explored the GaN NW synthesis based on InGaN/GaN MQW core-shell architecture NW arrays with a large area by MOCVD.[198] As a part of their study, Zhao et al. investigated the temperature-dependent and time-resolved carrier dynamics by conducting temperature-dependent device measurements of high-power InGaN/GaN quantum disks (Qdisks)-in-nanowire LEDs grown on molybdenum (Mo) capable of emitting at ~710 nm, thus extending beyond the true-green wavelength range.[133] These LEDs exhibited an ultralow turn-on voltage of ~2 V and were shown to be droop free at an unprecedentedly high input power density of 45 kW/cm^2 (obtained for an NW-LED of 20 μm diameter).[133]

Here, we will show that the GaN NWs grown on Si by PLD are used as growth templates for the fabrication of two devices. The first device is an n-GaN NW/p-GaN-based LED. The second one is the III-nitride MQW structure by plasma-Assisted MBE (PA-MBE) technique. Two kinds of InGaN/GaN MQWs were grown on PLD GaN NWs, whereby "In" composition, sample temperature, and number of QWs were varied.

6.2. LED structure based on PLD GaN NWs on p-GaN

6.2.1. LED Device Fabrication

LEDs based on III-nitride materials are widely utilized due to their structural and mechanical hardness and their efficient radiative recombination rates.[199, 200] As a part of the present study, lateral LED (LLED) was fabricated based on the PLD GaN NWs. This LED structure is based on GaN NWs grown on a p-GaN film. The structure consists of PLD GaN NWs grown on a p-type GaN layer doped by Mg (0.5 μm) grown on the c-sapphire wafer, as shown in Figure 6.1. Au/Ti contacts layers were used with GaN NWs and Au/Ni with the p-GaN substrate, as depicted in Figure 6.1.

6.2.2. I–V Characterization

The LED structure results presented in Figure 6.2 show that low current is produced at high voltage (15 V). Further optimization is needed, and it will be part of future work. However, the wavelength emitted from LED could not be measured as it produces very weak EL emission. Thus, in future studies, the reasons behind this current reduction will be explored, along with the means of overcoming them, such as using Si dopants in NW, to increase the light intensity emitted from such LED device. Other materials will also be used for contact layers to align the LED band energy.

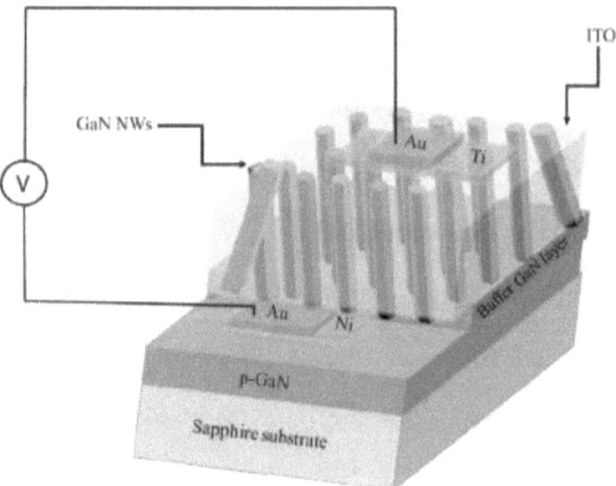

Figure 6.1. Schematic diagram of LLED structure.

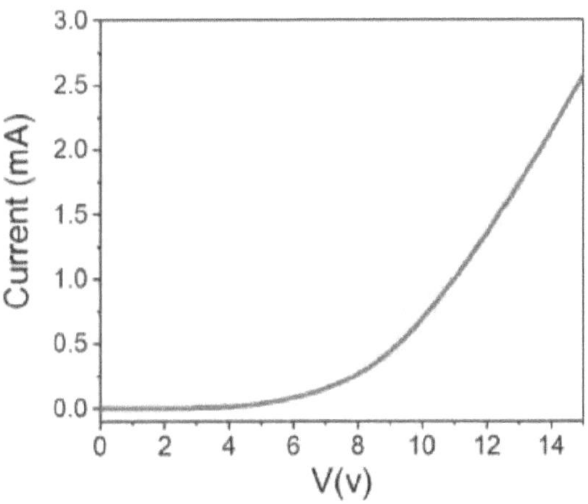

Figure 6.2. I-V curve of LLED grown on the p-GaN substrate.

6.3. LED structure based on InGaN/GaN MQW layers grown on PLD GaN NWs

The GaN NWs discussed here were grown on different bulk (including Si) and 2D substrates. GaN NWs were grown by using PLD, and the mechanism of GaN NW formation, along with the optimal growth conditions without a catalyst or seeding, was explored in detail in Section 4.3. For this purpose, PLD GaN NWs were positioned on an MBE holder (molybdenum block) before being placed in a load-lock chamber comprising the PA-MBE system to pre-outgas the samples at 200 ℃ for 1 hour to remove moisture. Next, the samples were transferred to the preparation chamber and were outgassed again at a higher temperature (650) ℃ for 2 hours. Finally, NWs were transferred into the MBE growth chamber and were mounted on the manipulator. Before growth, the substrate was heated to 800 ℃ for additional thermal cleaning to remove any residual organic contaminants.

By using the PA-MBE technique, GaN NWs/GaN layer/InGaN/GaN MQWs structure for use as an LED device was obtained, as shown in Figure 6.3a and 6.3b. Before the MQW growth on the GaN template, 500–100 nm thick GaN coalescent (buffer) layer was deposited. The MBE GaN layer was grown via a 2-step process. First, the sample was exposed to high Ga flux at 800 °C and N_2 plasma was irradiated on the surface of PLD-grown GaN nanowires to promote GaN layer growth. Initially, the Ga beam equivalent pressure (B.E.P.) for GaN growth was kept at 2.0×10^{-7} Torr, and plasma was maintained at 350 W power, with the nitrogen flow of 0.8 sccm. These growth conditions are considered Ga-rich conditions and are needed for GaN layer growth, as the gaps between GaN nanowires can be merged. After 5 hours, the growth temperature was decreased to 700 °C and the coalescent GaN layer was grown for 2 hours at a higher Ga B.E.P. (3.0×10^{-7} Torr) while maintaining the previous plasma (350 W) and nitrogen flow (0.8 sccm) settings.

The second set of InGaN/GaN MQWs was grown on the GaN buffer layer, prepared as described above. For this process, a B.E.P. of 1.0×10^{-7} Torr and 5.0×10^{-8} Torr was adopted for Ga and In, respectively, maintaining the N_2 flow of 0.8 sccm, while reducing the plasma power to 300 W. Based on the parameter settings, two groups (Set A and Set B) were grown, as shown Table 6.1.

Table 6.1. InGaN/GaN MQWs conditions.

Samples	The ratio between the "In" flux and the "Ga" concentration	Growth temperature	Number of QWs
Set A	5:5	650 °C	5
Set B	7:3	580 °C	15

6.3.1. Characterization

Figure 6.4a and the related inset show cross-sectional and top-view SEM images of the GaN buffer layer by MBE on PLD GaN NWs, while the STEM image of a GaN buffer layer on PLD NWs is provided in Figure 6.4b. In addition, the PL spectrum of a GaN buffer layer on PLD GaN NWs is depicted in Figure 6.5, showing sharp NBE GaN emission with no yellow band, indicating a good crystal quality.

Under the conditions adopted for Set A, no MQWs could be seen on the STEM images, as shown in Figure 6.6a and 6.6b, as such a small number of MQWs (5) is not observed by the TEM images. The PL spectrum of InGaN/GaN MQWs on PLD GaN NWs comprising Set A shows two UV emissions representing GaN at ~364 nm and InGaN at ~396 nm, as depicted in Figure 6.7.

Under the conditions adopted for Set B, 15 InGaN/GaN MQWs were grown on the NWs. Figure 6.8a shows the STEM images depicting the substrate/PLD GaN NWs/GaN buffer layer/InGaN/GaN MQWs structure, while Figure 6.8b shows the STEM image for InGaN/GaN MQWs based on the PLD GaN NWs template. The PL spectrum of InGaN/GaN MQWs on PLD NWs shows a UV peak at 365 nm and a dominant blue peak at 423 nm, representing GaN and InGaN emissions,[186, 190, 201] respectively, as depicted in

Figure 6.9. The PL emission of Set B is stronger than that of Set A due to the much greater number of MQWs in Set B.

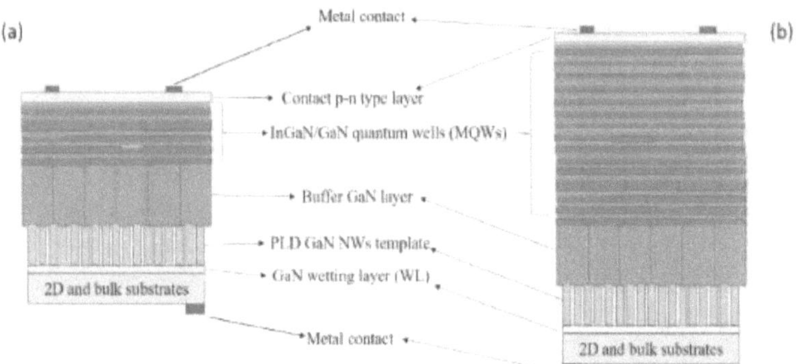

Figure 6.3. Schematic diagram of LED structure that used PLD GaN NWs as a template for (a) Set A and (b) Set B.

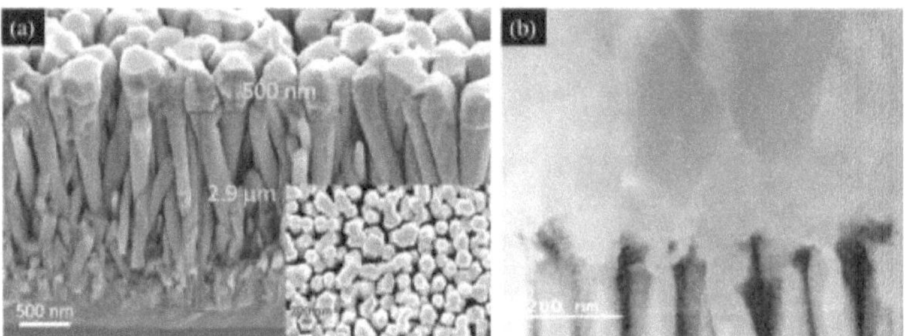

Figure 6.4. (a) Cross-sectional SEM image of GaN buffer layer on our GaN NWs and the inset shows top-view SEM. (b) STEM image for the GaN buffer layer on PLD NWs.

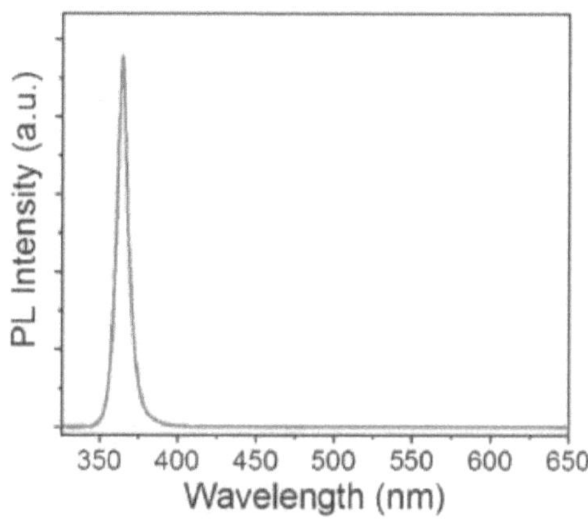

Figure 6.5. PL spectrum of GaN buffer layer after deposition on GaN NWs.

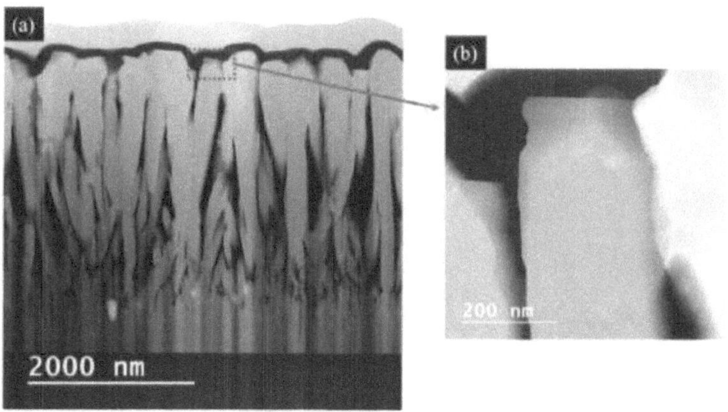

Figure 6.6. (a) STEM image for Set (A) of "5" InGaN/GaN MQWs based on PLD GaN NWs and (b) represents STEM image for the top of the same MQWs.

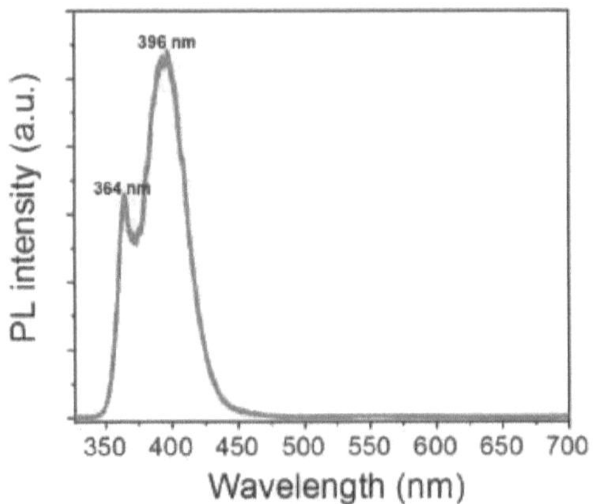

Figure 6.7. PL spectrum for InGaN/GaN MQWs (Set A) based on the PLD GaN NWs template.

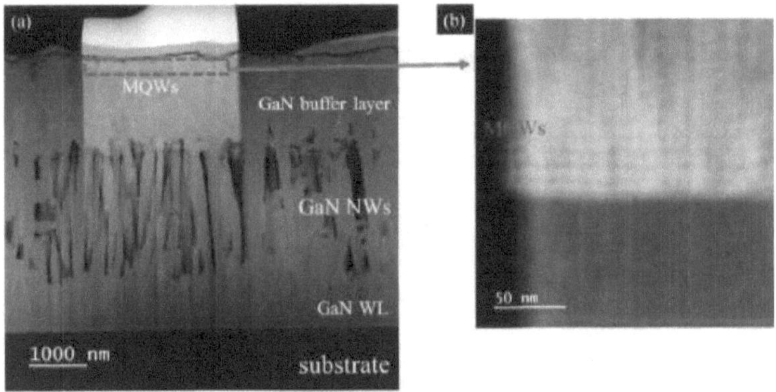

Figure 6.8. (a) STEM cross-sectional image for InGaN/GaN MQW structure grown on PLD GaN NWs (Set B) and (b) HR-STEM image of the 12 pairs of MQWs.

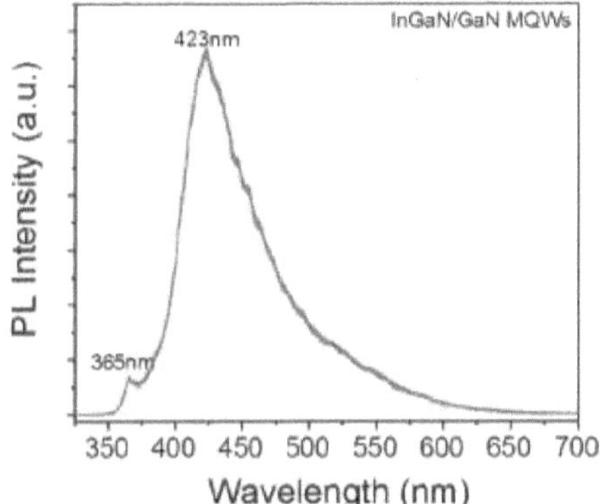

Figure 6.9. PL spectrum of InGaN/GaN MQWs (Set B) based on our PLD GaN NWs template.

6.4. Summary

As a part of this work, an LED structure based on PLD GaN NWs grown on the p-GaN substrate was fabricated. The low current was obtained from this LED, as this was the first attempt to obtain an LED-based on the newly developed GaN NWs. Moreover, GaN NWs grown by PLD on Si substrates were used as a growth template for device fabrication to grow III-nitride MQWs UV and visible-based LEDs and laser diodes (LDs). PA-MBE technique was used to grow InGaN/GaN MQWs layer on PLD NWs. The PL spectrum of InGaN/GaN MQWs on PLD NWs shows a UV peak at 365 nm, while a blue peak at 423 nm represents GaN and InGaN emissions.

Chapter 7

Comparative investigation: Enhanced Photodetectors based on GaN nanowires functionalized by different perovskites via work function engineering

7.1. Introduction

RT broadband photodetectors (PDs) are widely used in space science, chemical detection, optical communications, flame sensing, and national defense, among many other fields.[175, 202, 203] However, achieving high responsivity and adequate strength in the UV spectral range comparable to that obtained in the visible region of the electromagnetic spectrum remains a challenge. This issue has prompted researchers to consider combining solution-processed organic/inorganic perovskites with wide-bandgap semiconductors when fabricating high-performance devices at a relatively low cost, as this design has the potential to improve the detection in the UV range.[204]

Two types of inorganic-organic halide perovskite ($CH_3NH_3PbI_3$, methylammonium lead iodide [MAPI]) and inorganic perovskite (all-inorganic cesium lead halide [$CsPbBr_3$])— were used for photodetector applications due to their suitable direct bandgap, a wide absorption area, and high charge carrier mobility.[205-207] However, as mentioned in Chapter 2 Section 2.9.1, all inorganic $CsPbBr_3$ has better material stability than that of $CH_3NH_3PbI_3$. It is also characterized by higher luminescence intensity compared to $CH_3NH_3PbI_3$.

Extensive work has been conducted on heterojunction perovskite/ZnO-based photodetectors due to the advantage of heterojunction devices over homojunction or

Schottky devices.[202, 208, 209] In heterojunction structure, the main advantage is low applied fields as well no oxygen dependency compared to other structure based on Schottky contact UV detectors, that require a high applied field and depend on the oxygen adsorption/desorption process.[175, 210] However, as the interaction with oxygen in the atmospheric moisture and metal oxide materials leads to perovskite degradation, the device performance is suboptimal.[211, 212] Therefore, GaN has emerged as one of the best wide-bandgap semiconductors due to its good stability, as the weak oxidation of nitrogen may result in better stability when hybridized with perovskite materials[213, 214] compared to metal oxide. As GaN has a direct wide bandgap (~3.4 eV) at RT,[18] its use can enhance the response wavelength of perovskite-based PDs, extending it to the UV region. These materials also possess excellent thermal stability (2.1 W/cm K) and conductivity (2.1 W/cm K)[215], rendering them suitable for applications in optoelectronic high-power and high-frequency devices. . Thus, organic-inorganic halide perovskite $CH_3NH_3PbI_3$/GaN heterojunction based photodetector has been demonstrated.[100, 213] However, GaN layer was based on film structures. In addition, However, all-inorganic perovskite/GaN based devices have not been investigated yet. In particular, no perovskite/GaN nanowires (NWs) based devices has been studied.

GaN NWs have attracted significant attention from researchers and practitioners in recent years due to their superior physical properties compared to their bulk counterparts (microstructure materials).[216, 217] These features have prompted research into their potential usage in the fabrication of electronic and optoelectronic devices.[216]

In this chapter, PLD was successfully applied for GaN NW growth without a catalyst or seeding, using a p-type Si substrate as shown in Chapter 4. The obtained GaN NWs were subsequently used to develop a broadband photodetector by depositing all-inorganic $CsPbBr_3$ perovskites on GaN NWs as it has a good alignment with GaN materials. A photodetector based on organic-inorganic MAPI/GaN NWs has been fabricated for comparison. Further evaluations were performed to demonstrate that the photodetector works effectively in both UV and visible spectral range.

7.2. Experimental Method

The GaN NWs were grown on single-side-polished p-type Si (100) substrate using PLD, whereby the mechanism of GaN NW formation and the optimum conditions for growing GaN NWs without a catalyst or seeding are discussed in Chapter 4. As previously noted, two types of perovskite were used in device fabrication, one of which was $CH_3NH_3PbI_3$ perovskite in powder form (99% purity, purchased from Xi'an Polymer Light Technology Corp).[202] To obtain a uniform perovskite layer, 0.25 M (1 M = 0.619.9 g/mL) of $CH_3NH_3PbI_3$ perovskite powder was mixed with 1 mL of dimethylformamide (DMF) at 60 °C. Next, perovskite was spray-coated on GaN NWs using a compressed-air brush (pro series BD-132) producing nitrogen gas flow at 1 bar pressure perpendicular to the GaN NW sample, which was placed at approx. 15 cm above a hot plate, before being annealed for 10 minutes at 100 °C.[202] For the preparation of the second perovskite ($CsPbBr_3$), Lead (II) bromide ($PbBr_2$, 99.999% trace metals basis), cesium acetate (CsAc, 99.99% trace metals basis), octylamine (OcAm, 99%), octanoic acid (OcAc, 98%), 1-propanol (PrOH), n-hexane (Hex, 99%), and toluene (99.8%) were purchased from Sigma-Aldrich.[90] All chemical materials were utilized without any further purification. $CsPbBr_3$ NCs were synthesized by a modified procedure reported previously.[218] Typically, the Cs precursor and $PbBr_2$ precursor were prepared separately, and the reaction was initiated by injecting the latter into the former. First, Cs precursor solution was prepared by dissolving 32 mg of CsAc in 1 mL of 1-PrOH in a 20 mL vial under stirring in the air at room temperature, followed by the addition of 6 mL of Hex and 2 mL of 1-PrOH. Second, the $PbBr_2$ precursor solution was prepared by dissolving 245 mg of $PbBr_2$ into a mixture of 0.45 mL of 1-PrOH, OcAc, and OcAm each, which were combined at 90 °C in the air under vigorous stirring. Third, the hot $PbBr_2$ precursor was injected into the Cs precursor swiftly under vigorous stirring at room temperature. The system turned green immediately, and the reaction completed in 2 minutes. The $CsPbBr_3$ NCs were isolated by centrifugation at 3000 rpm of 4 min duration, and the pellet was dispersed into 2 mL of toluene.[90] Finally, $CsPbBr_3$ perovskite was drop-casted on GaN NWs.

To deposit the transparent indium-tin-oxide (ITO) electrode, the sample was transferred directly to a magnetron sputtering vacuum chamber. A 150 nm-thick ITO layer was deposited as a top electrode on both perovskite types by using a shadow mask. For this purpose, radio frequency magnetron sputtering at room temperature was used, with argon plasma at 5 mTorr working pressure and a constant current of 0.15 A for 100 minutes. Also, a gold layer (Au) was deposited to serve as the bottom contact with the p-Si substrate, while a silver (Ag) layer ensured top contact with GaN NWs.

NWs material morphology was characterized by scanning electron microscopy (SEM) (FEI Nova Nano 630). Besides, crystallization measurements were performed using X-ray diffraction (XRD) system (Bruker D8 Discover high-resolution XRD with Cu Kα and λ = 1.5406 Å). To measure the optical properties of the GaN NWs deposited on a p-Si substrate, RT PL measurements using Horiba LabRAM Aramis with He-Cd laser (of λ = 325 nm) was performed. PL setup was attached to the monochromator and CCD camera. RT absorption measurements were carried out using UV-vis Varian Cary 5000 spectrophotometer to evaluate the bandgap of both GaN NWs and perovskites, and to explore the produced spectra, aiming to identify any absorption edges that could influence the device performance. RT I−V characteristics were measured using a Keithley 4200 analyzer and a light-emitting diode with a light power of 53 mW cm^{-2}. Photoresponsivity (R) and detectivity (D^*) were calculated using equations 2.20 and 2.21 in Chapter 2.

7.3. Results and Discussion

7.3.1. Materials Properties

Figure 7.1a show tilted-view SEM images of the GaN samples without perovskite layer, whereby Figure 7.1b and 7.1c respectively show a 1500 nm thick $CsPbBr_3$ and $CH_3NH_3PbI_3$ perovskite layers deposited on the GaN NWs,[90, 202] indicating that the perovskite has filled the gaps between the GaN NWs.

To investigate the crystalline structure of GaN NW/ perovskite photodetectors developed as a part of the present study, XRD 2θ scans were obtained. Figure 7.2a shows X-ray diffraction patterns of GaN NWs/ $CsPbBr_3$ perovskite sample, while the XRD peaks of GaN NWs/$CH_3NH_3PbI_3$ perovskite are shown in Figure 7.2b. The 002 peak at ~35° is related to GaN NWs as shown in Chapter 4 Section 4.3.4. The 15.6°, 21.8°, 31°, 38.1° and 44.2° peaks are attributed to plans 110, 020, 004, 312 and 224 of $CsPbBr_3$, whereas 14.8°, 20.6°, 24.17°, 29.1°, 32.5°, 41.1°, 43.64°, 48°, 50.8°, 55.1°, 57° and 59.4° peaks are related to 110, 200, 202, 220, 213, 400, 330, 413, 404, 424, 433 and 440 planes in CH3NH3PbI3. The XRD findings are in good agreement with the previously reported experimental results related to both perovskite types.[219, 220]

PL measurements were also performed, to investigate the material quality.[221] The RT PL spectra obtained for the GaN NWs before perovskite deposition indicate the presence of a sharp dominant GaN band edge emission at 350 nm when excited by a 325 nm laser line, along with a very weak defect peak (yellow band), as shown in Figure 7.3a. Figure 7.3b and 7.3c show the PL spectra of both perovskite materials deposited on the glass when excited by 475 nm line. PL spectrum of perovskite $CsPbBr_3$ deposited on glass shown in Figure 7.3b reveals a strong peak at 520 nm, which in agreement with previous literature[222] and the PL measurements confirm that a strong emission peak at 769 nm arises from the $CH_3NH_3PbI_3$ perovskite,[221] The absorption results illustrate the working range of the photodetector developed as a part of this study. Figure 7.4a and 7.4b show the absorption results related to p-Si substrate/ GaN NWs/ $CsPbBr_3$ perovskite and p-Si substrate/ GaN NWs/ $CH_3NH_3PbI_3$ perovskite, respectively, confirming that the photodetector working range has been extended from the visible part of the spectrum to include the UV region. This enhancement is due to the presence of GaN, as previously reported by other authors.[90, 213]

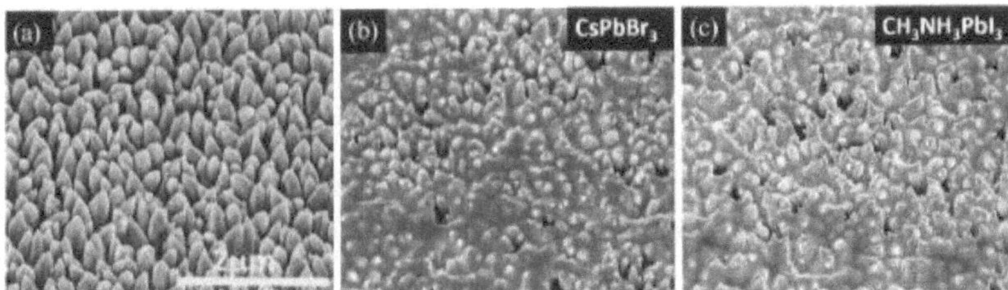

Figure 7.1. SEM images of (a) bare GaN NWs, (b) GaN NWs/ $CsPbBr_3$ perovskite, and (c) GaN NWs/ $CH_3NH_3PbI_3$ perovskite.

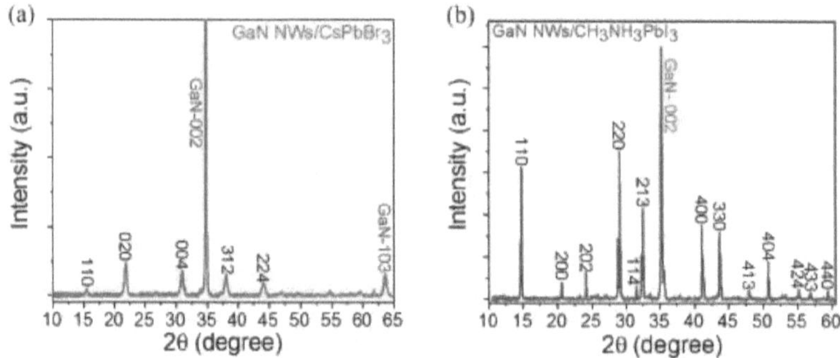

Figure 7.2. XRD patterns of (a) GaN NWs/CsPbBr$_3$ perovskite and (b) GaN NWs/ CH$_3$NH$_3$PbI$_3$ perovskite.

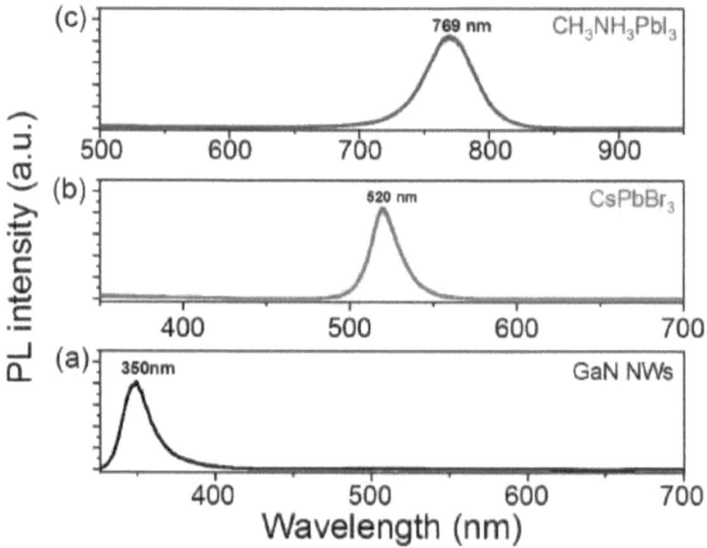

Figure 7.3. PL results of (a) GaN NWs, (b) CsPbBr$_3$ perovskite on glass and (c) CH$_3$NH$_3$PbI$_3$ perovskite on glass.

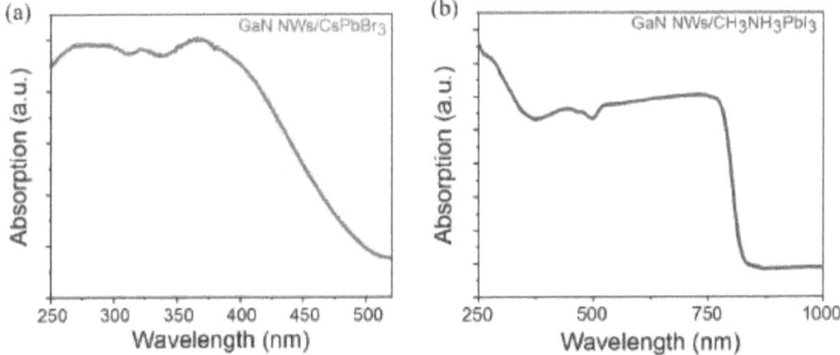

Figure 7.4. Absorption results of (a) GaN NWs/ CsPbBr₃ perovskite and (b) GaN NWs/ CH₃NH₃PbI₃ perovskite.

7.3.2. Device characteristics

Given that both top–back contact (vertical) configuration and top–top contact (lateral) configuration have been used for both the perovskite materials in the present study, in the sections below, device characteristics and performance will be discussed in the sections below.

i) *Si/ GaN NWs/perovskite vertical photodetector device configurations*

The schematic diagram of the vertical photodetector configurations based on Au/p-Si/GaN NWs/ CsPbBr₃/ ITO and Au/p-Si/GaN NWs/ CH₃NH₃PbI₃/ ITO structures are shown in Figure 7.5a and 7.5b, respectively, whereby Figure 7.6a and 7.6b provide the relevant energy levels of the materials used in these photodetectors, respectively. The valence band (VB) and the conduction band (CB) of n-GaN are -7.1 and -3.7 eV, respectively.[223] The VB and CB of CH₃NH₃PbI₃ perovskite are -5.4 and -3.9 eV,[224] while -5.6 eV and -3.3 eV are reported for the CsPbBr₃ perovskite, respectively.[225] For Si, the VB is -5.17 and the CB is -4.05 eV, as shown in Figure 7.6a and 7.6b.[226] Under white light illumination, a positive/negative bias was applied to the anode (ITO)/cathode (Au), due to which photogenerated carriers were created in the GaN NWs/perovskite photodetector. In the photodetector based on GaN NWs/CsPbBr$_3$ perovskite, there is a good band alignment between GaN NWs and CsPbBr$_3$ perovskite, due to which the electrons can transition from the perovskite layer to the GaN layer, passing through the Si substrate before being collected by the Au electrode, as shown in Figure 7.6a. On the other hand, in the device based on GaN NWs/CH$_3$NH$_3$PbI$_3$ perovskite, the electrons can easily overcome the barrier between the CH$_3$NH$_3$PbI$_3$ perovskite and the GaN layer, passing through the Si substrate before being collected by the Au electrode, as shown in Figure 7.6b.

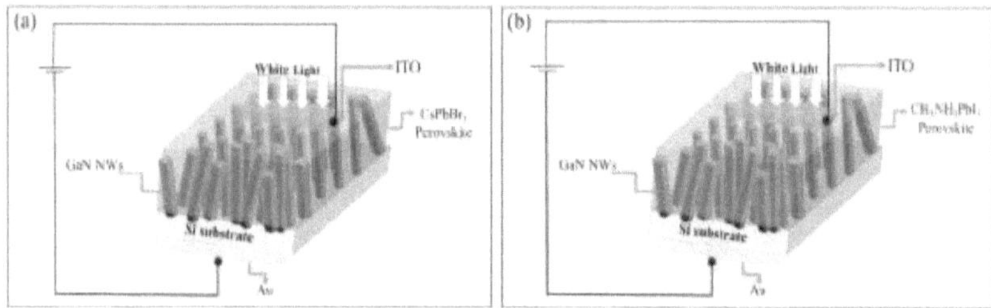

Figure 7.5. Schematic diagram of photodetectors (with vertical configuration) based on (a) CsPbBr$_3$ perovskite and (b) CH$_3$NH$_3$PbI$_3$perovskite.

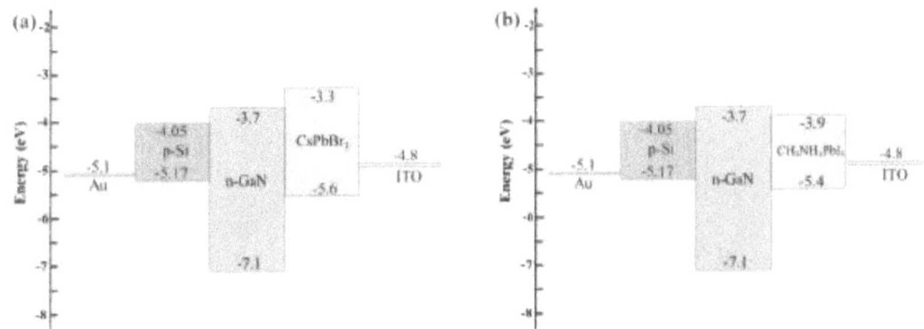

Figure 7.6. Energy band diagram of the photodetector with vertical configuration based on (a) CsPbBr3 perovskite and (b) CH3NH3PbI3 perovskite.

To study the response of the photodetectors, the I−V characteristics were measured under white light illumination provided by a 53 mW/ cm^2 light source. To study the response of the photodetectors, the I−V characteristics were measured under white light illumination provided by a 53 mW/ cm^2 light source. In the photodetector based on p-Si/ GaN NWs/ CsPbBr3 perovskite, within the -0.15−0.15 V range, the current increased from 0.38 μA to 0.8 μA, and from 0.003 μA to 0.009 μA at 0 V, as shown in Figure 7.7a. For the device based on the p-Si/ GaN NWs/CH3NH3PbI3 perovskite, at an applied bias in the -5−5 V range, the dark current was 0.5 μA, and the photocurrent increased to 2.2 μA when the device was illuminated, and from 0.0005 μA to 0.008 μA at 0 V, as shown in Figure 7.7b. In each case, the curves related to the dark current and photocurrent indicate good rectifying behavior. The maximum voltage at which the CsPbBr3 perovskite device operated was 0.15 V, indicating that the small turn-on voltage compared to that based on GaN NWs/CH3NH3PbI3 perovskite. As can be seen in Figure 7.7a, unlike GaN NWs/CH3NH3PbI3 a slight shift in the current is produced under illumination in the GaN NWs/ CsPbBr3 photodetector, confirming self-powered behavior.

Photoresponsivity (R) and detectivity (D*) were also calculated, as these are the key parameters for estimating photodetector performance. Under white light illumination at 0.1 V, R = 37.7 mA/W was measured for p-Si/ GaN NWs/CsPbBr$_3$ perovskite photodetector and D* = 2.21 × 10^{13} Jones was obtained. However, for comparison, the R-value is 18.5 mA/W at 0.1 V for GaN NWs/ CH$_3$NH$_3$PbI$_3$ photodetector, which is much smaller. It is noteworthy that this vertical Si/GaN NWs/CsPbBr$_3$ photodetector is as a self-powered device under white light illumination at 0 V, due to a good band alignment without barriers blocking the carrier flow GaN (source of electron carriers) and CsPbBr$_3$ perovskite (source of hole carriers), unlike the band alignment between CH$_3$NH$_3$PbI$_3$ and GaN that causes some carriers to be blocked from CH$_3$NH$_3$PbI$_3$ to GaN due to the higher work function level of GaN, as shown in Figure 7.6a and b. The internal electric field created in the depletion layer between GaN (which generates electrons) and CsPbBr$_3$ perovskite (which generates holes) is responsible for electron−hole separation even at 0 V, creating photogenerated carriers and transient response,[213, 227] resulting in R = 147.8 mA/W and D* = 7.62 × 10^{14} Jones, which is much higher than that of GaN NWs/CH$_3$NH$_3$PbI$_3$ reported in this work and previous literature.[100, 213] This is due to the good band alignment between GaN and CsPbBr$_3$ compared to that of CH$_3$NH$_3$PbI$_3$. In addition, no self-powered characteristics have been shown in Schottky CsPbBr$_3$-based photodetector as reported previously, further confirming the advantages of heterostructure configurations in our devices over the Schottky devices.[90]

On the other hand, for the GaN NWs/CH$_3$NH$_3$PbI$_3$ photodetector, at 5 V, R = 94 mA/W was obtained under illumination. These values are remarkable considering that this is the first photodetector based on GaN NWs/CH$_3$NH$_3$PbI$_3$ perovskite, as well as when compared

143

with the results reported for CH$_3$NH$_3$PbI$_3$ perovskite/GaN film (26 mA/W and 55 mA/W).[100, 213] The high photoresponsivity can be referred to the large surface-volume ratio for NWs compared to film, allowing large light exposure. A detectivity of D* = 7.22 × 10^{13} Jones was obtained for GaN NWs/CH$_3$NH$_3$PbI$_3$ photodetector, which is comparable to the values reported for perovskite photodetectors.[213]

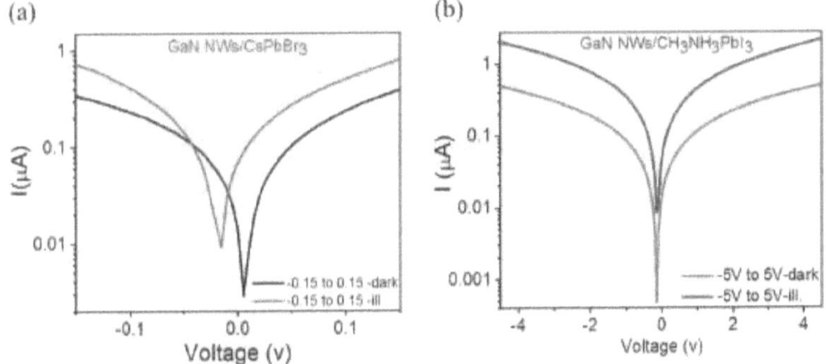

Figure 7.7. I–V measurements of vertical photodetectors based on (a) CsPbBr$_3$ perovskite, and (b) CH$_3$NH$_3$PbI$_3$ perovskite, under white light illumination at 0.53 mW/cm^2.

For the vertical configuration photodetector based on p-Si/GaN NWs/CsPbBr$_3$ perovskite, the wavelength-dependent measurements shown in Figure 7.8a, reveals that no response in the UV region was obtained, whereas, under visible light region illumination, there is much better photoresponse compared than that from p-Si/GaN NWs/CH$_3$NH$_3$PbI$_3$ photodetector, due to a good band alignment between GaN and CsPbBr$_3$ perovskite. As shown in Figure 7.6a, the carriers are generated from GaN (electrons) and perovskite (holes) forming a heterojunction active layer based on these two materials, whereas Si acts as the carrier transport layer in p-Si/GaN NWs/CsPbBr$_3$ device. Thus, when GaN absorbs UV radiation, charge transfer occurs from GaN to $CsPbBr_3$ perovskite due to strong band alignment. In this case, fast carrier recombination in the perovskite materials that is faster than the transient time that required to transport the carriers to anode/cathode.[213, 227] Figure 7.8b shows the transient characteristics of the vertical p-Si/ GaN NWs/ $CsPbBr_3$ perovskite photodetector under white illumination and 0.1 V demonstrating the photoresponse the device using several on-off cycles. Figure 7.8c shows transient characteristics of this photodetector under white light illumination at 0 V, further confirming that it can operate as a self-powered device.

While for the photodetector based on GaN NWs/$CH_3NH_3PbI_3$, the deep UV photoresponse at 200 nm is enhanced due to the GaN NW incorporation (R = ~53 mA/W), as shown in the photoresponsivity curve as a function of different wavelength lines (Figure 7.9a). This finding is also in good agreement with the responsivity results obtained for the $CH_3NH_3PbI_3$ perovskite/GaN film photodetector.[213] In this case, as shown in Figure 7.6b, the barrier between $CH_3NH_3PbI_3$ perovskite and GaN, whereas the band alignment between p-Si and n-GaN NWs is good, resulting in a p−n junction between Si and GaN which serves as an active layer and the $CH_3NH_3PbI_3$ perovskite acts as the hole transport layer under UV illumination. On the other hand, under visible light illumination (in particular higher R-value (146.8 mA/W) is observed at 800 nm compared to other wavelengths), the $CH_3NH_3PbI_3$ perovskite acts as the active layer and GaN and Si serve as the carrier transport layer, which in line with previous reports.[100, 213]

Figure 7.9b shows the transient characteristics of the vertical configuration p-Si/GaN NWs/$CH_3NH_3PbI_3$ perovskite photodetector under white light illumination at 5 V, using several on/off cycles. For this photodetector, the self-powered behavior is not observed due to the barrier between GaN and $CH_3NH_3PbI_3$ perovskite, which prevents p–n junction formation between these two layers. As explained above, each layer acts independently, as the active and the transport layer for their respective wavelengths.[228]

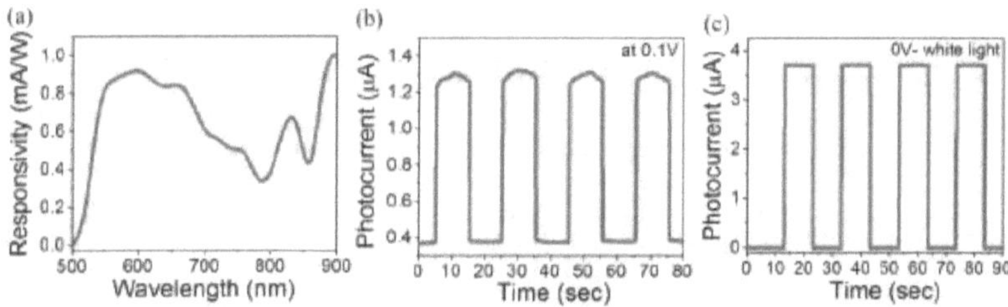

Figure 7.8. (a) Vertical p-Si/ GaN NWs/ CsPbBr$_3$ photodetector responsivity as a function of wavelength, and photoresponse under white light at (b) 0.1 V and (c) 0 V.

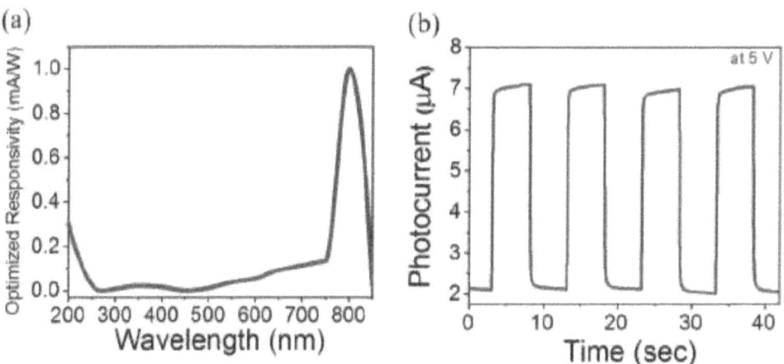

Figure 7.9. (a) p-Si/ GaN NWs/ CH$_3$NH$_3$PbI$_3$ vertical photodetector responsivity as a function of wavelength and (b) transient response under white light at 5 V.

ii) ***GaN NWs/perovskite lateral photodetector device configurations***

In the lateral configuration photodetectors, the Au back-layer contact was replaced by an Ag layer, serving as the top contact on GaN NWs, while the ITO layer provides top contact on the region that perovskite is hybridized with GaN NWs, as shown in Figure 7.10a and

7.10b. The schematic diagram of a lateral configuration photodetector, of Ag/GaN NWs/ CsPbBr$_3$ perovskites/ITO, is shown in Figure 7.10a, while 7.10b shows the structure of Ag/GaN NWs/ CH$_3$NH$_3$PbI$_3$ perovskites/ITO. The selection of electrodes is based on previous studies.[100] Under white light illumination, a positive/negative bias was applied to the anode (ITO)/ cathode (Ag), respectively, creating photogenerated carriers. Figure 7.11a and 7.11b depict the energy band alignment of the lateral configuration photodetectors based on CsPbBr$_3$ and CH$_3$NH$_3$PbI$_3$ perovskites. Note that p-Si is not included here as Si here is acted as the device platform only. As was the case for the vertical configuration photodetectors, in those based on the photodetector based on CsPbBr$_3$ perovskite, there is a good band alignment between GaN NWs and CsPbBr$_3$ perovskite, allowing the electrons to move from the perovskite layer to the GaN layer to be efficiently collected by the electrode, as shown in Figure 6.11a compared to that based on CH$_3$NH$_3$PbI$_3$. On the other hand, CH$_3$NH$_3$PbI$_3$ perovskite, the electrons can easily overcome the barrier between the CH$_3$NH$_3$PbI$_3$ perovskite and the GaN layer and are collected by the electrodes, as shown in Figure 7.11b.

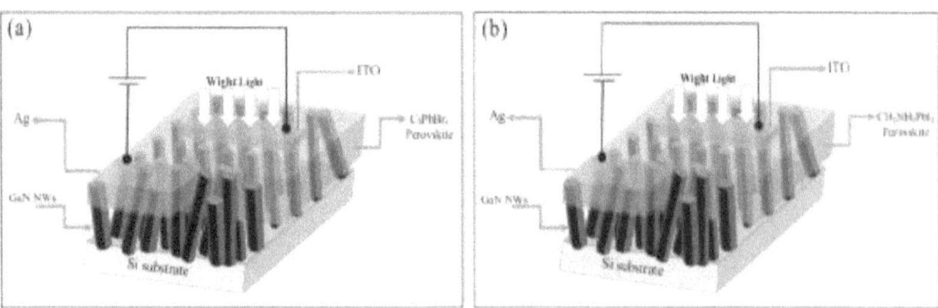

Figure 7.8. Schematic diagram of a photodetector with lateral contact configuration based on (a) GaN NWs/ CsPbBr$_3$ perovskite and (b) GaN NWs/ CH$_3$NH$_3$PbI$_3$ perovskite.

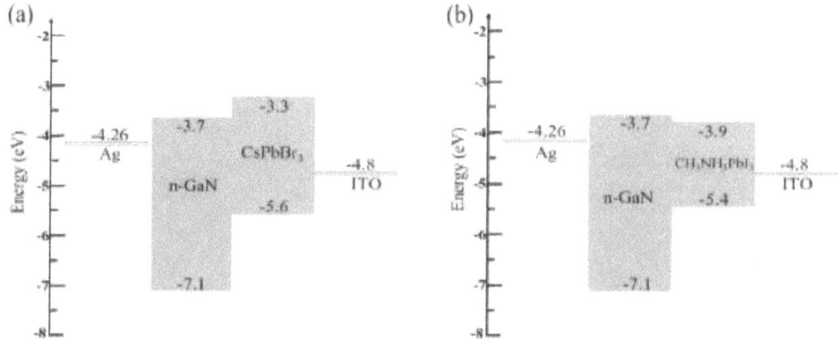

Figure 7.9. Energy band diagram of a photodetector with lateral configuration based on (a) GaN NWs/ CsPbBr$_3$ perovskite and (b) GaN NWs/ CH$_3$NH$_3$PbI$_3$ perovskite.

To study the photoresponse of the lateral photodetectors, the I–V characteristics were measured under dark and white light illumination provided by the same 53 mW/cm^2 light source used for vertical devices. For the lateral configuration photodetector based on Ag/GaN NWs/CsPbBr$_3$/ITO, at -0.15 to 0.15 V as a maximum operating voltage, the current increased from 1.7 μA to 2.9 μA, and from 0.004 μA to 0.02 μA at 0 V, as shown in Figure 7.12a. This indicates that the behavior of the lateral configuration Ag/GaN NWs/CsPbBr$_3$/ITO photodetector is very similar to that exhibited by the vertical GaN NWs/CsPbBr$_3$ configuration photodetector because there is a slight shift in the current under illumination, as shown in Figure 7.12a, confirming self-powered behavior, indicating photocarriers generation at 0 V.

For the Ag/GaN NWs/CH$_3$NH$_3$PbI$_3$/ITO device, at an applied bias between -5 and 5 V, the dark current was 0.14 μA, and the photocurrent increased to 0.7 μA when the device was illuminated, as shown in Figure 7.12b, exhibiting a weaker photocurrent compared to the vertical device (7.4 μA under illumination), suggesting that in vertical configuration device, photocarrier contusion from Si enhances the photoresponse of the GaN NW/ $CH_3NH_3PbI_3$-based devices.

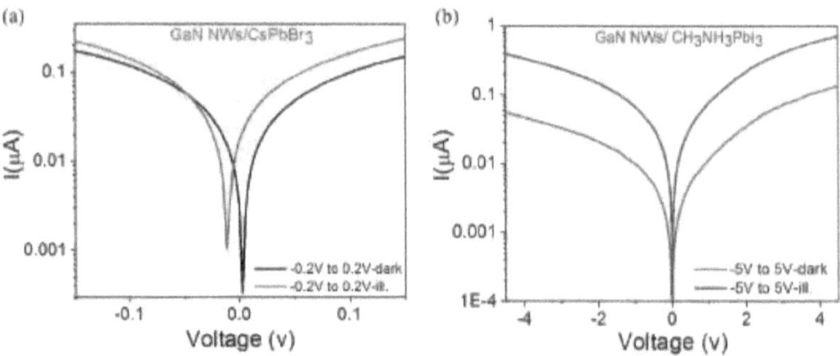

Figure 7.10. I−V measurements of lateral photodetectors based on (a) $CsPbBr_3$ perovskite, and (b) $CH_3NH_3PbI_3$ perovskite, under white light illumination at 0.53 mW/ cm^2.

For the lateral configuration photodetector based on GaN NWs/$CsPbBr_3$ perovskite, the wavelength-dependent measurements shown in Figure 7.13a, reveals that no response in the UV region was obtained, whereas, under visible light region illumination, there is much better photoresponse compared than that from p-Si/GaN NWs/$CH_3NH_3PbI_3$ photodetector, due to a good band alignment between GaN and $CsPbBr_3$ perovskite. For the lateral Ag/GaN NWs/$CsPbBr_3$/ITO photodetector, R = 17.4 mA/W and D* = 1.57×10^{13} Jones were measured under white light illumination at 0.1 V, as shown in Figure 7.13b, indicating that the vertical configurations is more effective (R = 37.7 mA/W) than lateral configurations.

Besides, Figure 7.13c shows the transient photoresponse characteristics of the lateral GaN NWs/ $CsPbBr_3$ photodetector under white light illumination at 0 V, confirming that it can operate as a self-powered device due to a good band alignment without barrier blocking (see Figure 7.11a), as was the case for the vertical GaN NWs/ $CsPbBr_3$ photodetector. In this photodetector, the same behavior was observed, whereby an internal electric field was created in the between GaN (that generates electrons) and $CsPbBr_3$ perovskite (that generates holes), which was responsible for electron–hole separation even at 0 V, creating photogenerated carriers and transient response,[213, 227] which results in R = 105 mA/W and $D^* = 1.2 \times 10^{14}$ Jones. Surprisingly, unlike vertical GaN NWs/ $CsPbBr_3$, this lateral configuration photodetector exhibited self-powered (0V) photoresponse under 325 nm laser illumination, as shown in Figure 7.13d, yielding R = 3.84 mA/W and $D^* = 2.17 \times 10^{12}$ Jones.

For the GaN NWs/ $CH_3NH_3PbI_3$ photodetector, under the deep UV (200 nm) illumination, the R-value is calculated to be 32 mA/W, as shown in Figure 7.14a. However, under white illumination at 5 V, a large R-value (82.2 mA/W) was obtained, as shown in Figure 7.14b, while illumination under 800 nm the R-value reaches the maximum (80.2 mA/W). In general, the R values of lateral GaN NWs/ $CH_3NH_3PbI_3$ devices are smaller values than those obtained by the vertical counterpart due to weaker photocurrent produced by lateral devices. As noted previously, still these values are considerable when compared with $CH_3NH_3PbI_3$ perovskite/GaN film results.[213] Moreover, the detectivity value of $D^* = 3.78 \times 10^{13}$ Jones obtained for this photodetector is comparable to the values reported for perovskite photodetectors.[213]

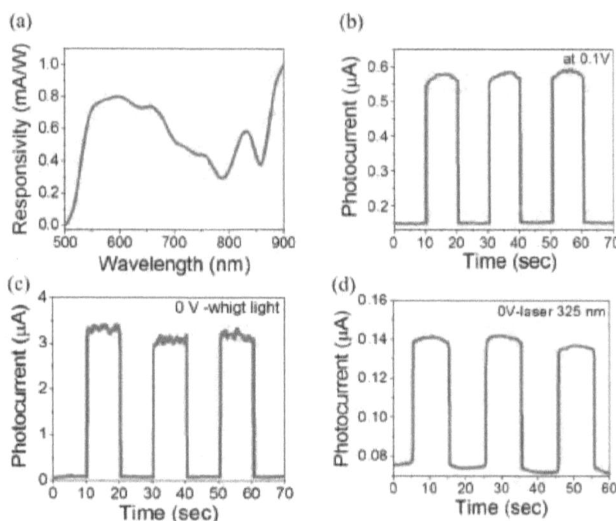

Figure 7.13. (a) Lateral GaN NWs/ CsPbBr₃ photodetector responsivity as a function of wavelength. Photoresponse of the lateral configuration Ag/ GaN NWs/ CsPbBr₃/ ITO photodetector under (b) white light at 0.1 V, (c) white light at 0 V, and (d) laser 325 nm illumination.

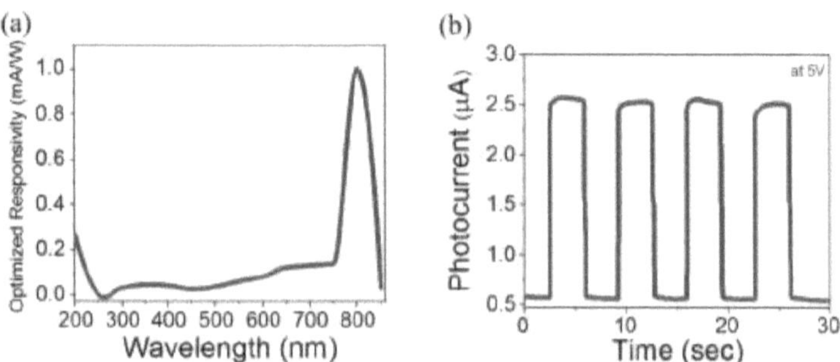

Figure 7.4. (a) Lateral GaN NWs/ CH₃NH₃PbI₃ photodetector responsivity as a function of wavelength and (b) transient photoresponse under white light at 5 V.

Table 7. 1. The comparison of all parameters (R, D*) for all four vertical and lateral devices in this work and previously reported Works.

samples	R (mA/W)	D* (Jones)	Self-powered	References
Vertical configuration p-Si/GaN NWs/$CH_3NH_3PbI_3$	R=94 at 5V R= 18.5 at 0.1V	7.22×10^{13}	No	This work
Lateral configuration p-Si/GaN NWs/$CH_3NH_3PbI_3$	R=82.2 at 5V R=10.8 at 0.1V	3.78×10^{13}	No	This work
Vertical configuration p-Si/GaN NWs/$CsPbBr_3$	R=37.7 at 0.1V	2.21×10^{13}	Yes, R=147.8 $D*=7.62\times10^{13}$	This work
Lateral configuration p-Si/GaN NWs/$CsPbBr_3$	R=17.4 at 0.1V	1.57×10^{13}	Yes, R=105 $D*=1.2 \times 10^{14}$	This work
Other work on GaN film/$CH_3NH_3PbI_3$	R=26 at 0.2V	2.89×10^{12}	Yes, R=160 $D*=7.96\times 10^{12}$	Ref. 211
Other works on GaN film/$CH_3NH_3PbI_3$	R=55 at 5V	———	No	Ref. 100

All devices the transient times (both rise (τ_r) and fall (τ_f) time) are calculated to be faster than 80 ms, which is the Keithley 4200 detection limit used in the work. Table 7.1 shows the comparison of all parameters (R, D*) for all four vertical and lateral devices. We found that vertical device configuration showed better performance, however, UV photoresponse for lateral devices based on GaN NWs/$CsPbBr_3$ is obtained.

7.4. Summary

In summary, as a part of this work, a RT broadband UV/visible photodetector based on the GaN NWs/ organic or inorganic perovskite was successfully developed using p-Si substrates for the first time. GaN NWs were grown by a direct method (PLD technique) without the need for costly and complex fabrication processing and without a catalyst or seeding. Two types of halide perovskites were used, inorganic-organic ($CH_3NH_3PbI_3$) and all inorganic ($CsPbBr_3$) perovskite. In this work, four photodetectors were discussed based on the contact layer design. First, the vertical p-Si/ GaN NWs/ $CsPbBr_3$ photodetector yielded R = 37.7 mA/W and D* = 2.21×10^{13} Jones under illumination at 0.1 V. In addition, this photodetector was shown to work as a self-powered device at 0 V, where the values of R = 147.8 mA/W and D* = 7.62×10^{14} Jones were obtained. Second, it was shown that the vertical p-Si/ GaN NWs/ $CH_3NH_3PbI_3$ photodetector exhibited good responsivity in the visible and UV regions, yielding R = 94 mA/W and D* = 7.22×10^{13} Jones. Its deep UV photoresponse at 200 nm was enhanced to 53 mA/W. Third, for the lateral GaN NWs/ $CsPbBr_3$ photodetector, R = 17.4 mA/W and D* = 1.57×10^{13} Jones were measured under white light illumination at 0.1 V. This device could operate in self-powered mode at 0 V, yielding R = 105 mA/W and D* = 1.2×10^{14} Jones. Under 325 nm laser illumination, photoresponse was noted, with R = 3.84 mA/W and D* = 2.17×10^{12} Jones. Finally, for the lateral GaN NWs/ $CH_3NH_3PbI_3$ photodetector, R = 82.2 mA/W and D*= 3.78×10^{13} Jones were obtained under illumination at 5 V and its deep UV photoresponse at 200 nm was enhanced to 32 mA/W. Despite low UV responsivity, these results are encouraging, given that this is the first attempt to develop a photodetector based on GaN NWs/ organic or inorganic perovskite architecture.

Chapter 8

Enhanced UV efficiency of GaN nanowires functionalized by wider bandgap solution-processed p-MnO quantum dots via the Energy transfer process

8.1. Introduction

Technologies based on GaN and related III-nitrides are widely used in several applications, such as light-emitting diodes (LEDs), laser diodes (LDs), photodetectors, and high-power and high-frequency devices,[229, 230] owing to the unique electrical and optical properties of these materials, such as direct bandgap, wide spectral tunability, good conductivity, and durability in harsh environments. In particular, GaN-based UV LEDs have been extensively studied, with the view of utilizing them in curing, medical devices, and diverse industrial domains.[231-233] However, the UV GaN-based LED performance is significantly below that of the visible emitting devices, due primarily to low external efficiency (< 10%) at the emission wavelengths below 365 nm.[234, 235] The low internal quantum efficiency (IQE) is one of the issues that reduce the external efficiency as a result of high threading dislocation (TD) density caused by lattice mismatch with commonly utilized substrates, which introduces high density of non-radiative centers.[233, 234] Therefore, developing a UV LED structure that can yield the required efficiency remains highly challenging.

Recently, GaN nanowires (NWs) have emerged as promising candidates for enhancing the optical efficiency of UV LEDs and reducing the effects of TDs. Indeed, their large surface-to-volume ratio promotes an elastic strain relaxation mechanism preventing the formation of dislocations.[235] Moreover, NW structures, which are inherently diffusive, permit light extraction efficiency enhancement.[236] However, the surface states characterizing NWs may lead to non-radiative recombination, possibly reducing optical efficiency significantly. In this case, Fermi level pinning effects, along with an enhanced surface non-radiative rate will reduce LED efficiency due to the Shockley−Read−Hall (SRH) non-radiative recombination at these surface states (defect trapping centers).[233, 237] This issue was partly mitigated by NW surface passivation using complex chemicals, which reduces the density of the surface non-radiative recombination.[233, 238] However, in this method, highly toxic solutions, such as diluted potassium hydroxide (KOH) solution, are used to remove unwanted surface defects from III-nitride nanowires.[233] Available evidence indicates that inhalation of vapor produced by such solutions can cause serious damage to the upper respiratory tract and the mucous membranes.[233,239] This method reduces the trap states only, resulting in insufficient enhancement (less than two folds enhancement).[233] Also, when the KOH passivation methods are used, <100% improvement in the UV emission of GaN NWs can be attained. Thus, there is still a need for a method that is environmentally friendly and is capable of producing greater improvements in the GaN-based NW UV emission, while being cost-effective for large-scale applications.[233, 239] This method reduces the trap states only, resulting in insufficient enhancement (less than two folds enhancement).[233] Thus, as safer and more effective alternatives are needed, significant efforts are still actively dedicated to the improvement of GaN emission, which would in turn result in more efficient UV LEDs.[240, 241]

QD-functionalized Si NW for solar cell applications has been used to enhanced light absorption, however, such design has not been used to enhance the mission emitted from III-nitride-based emitting devices.[242] Recently, we explored new wide bandgap (> 4.5 eV) crystalline p-type manganese oxide quantum dots (p-MnO QDs) synthesized by solution-processed femtosecond laser ablation technique in liquid (FLAL).[95] However, GaN NWs hybridized with wider-bandgap MnO (~ 4.8 eV, which is in deep UV spectral region) have not yet been adopted in devices to improve UV emission, as the effects of this structure on enhancing the GaN emission have not been investigated.

In this work was reported in this book, we demonstrate that our strategy based on functionalizing GaN NWs by solution-process p-MnO QDs is effective in enhancing the UV optical efficiency. This novel strategy is based on hybridizing p-type MnO QDs with GaN by a simple and cost-effective drop-casting method to improve the optical GaN efficiency through energy transfer from QDs to GaN. The obtained findings demonstrate that functionalizing GaN NWs by solution-processed p-MnO QDs is effective in enhancing the UV optical efficiency. This study included advanced optical characterizations and analyses, as well as electron energy loss spectroscopy (EELS) measurements, all of which indicated that the energy transfer (ET) can occur from the QDs to GaN, resulting in significant UV emission enhancement. Hence, this investigation provides valuable insight into the UV emitting device development. The novelty of our work stems from using wider-bandgap QDs to enhance the III-nitride UV and deep UV emission significantly, which is a method that has not been investigated. In addition, the optical properties of solution processed p-MnO QD-functionalized III-nitrides have not been investigated as well.

8.2. Experimental Method

8.2.1. p-type MnO QDs Synthesis

Our previous work on the properties of solution-processed p-type MnO QDs and their synthesis was reported elsewhere.[95] As a part of this previous investigation, MnO QDs were synthesized by cost-effective FLAL at room temperature (RT) and under ambient conditions (see Chapter 3 Section 3.1.2).[111] The FLAL synthesis of MnO QDs was carried out using Ti:sapphire Coherent Mira 900 laser (pulse width ~150 fs, 800 nm wavelength, operating at 76 MHz frequency) that was focused on the MnO target in ethanol solution.[95]

8.2.2. GaN NWs Synthesis

The GaN NWs were grown along the c-axis by catalyst-free, plasma-assisted molecular beam epitaxy (PA-MBE) in a MECA2000 MBE system was utilized that is in Prof. Bruno Daudin lab in University of Grenoble, France. The substrates were 2-inch wide Si (111) oriented wafers, which were de-oxidized in hydrofluoric acid prior to their introduction into the MBE chamber. The NWs were grown under nitrogen-rich conditions with a radio frequency power of 300 W for the plasma cell and a N_2 flux of 0.6 standard cubic centimeters per minute (sccm). The Ga/active N_2 flux ratio was typically equal to 0.3. A thin AlN buffer layer of around 3 nm was first deposited on Si prior to the growth of GaN NWs to prevent the in-plane tilt and twist of the wires and to homogenize the NW density across the entire Si wafer surface.

8.2.3. Structural Characterizations

NW material morphology was characterized by scanning electron microscopy (SEM, FEI Nova Nano 630). Moreover, high-resolution transmission electron microscopy (HR-TEM) and scanning TEM (STEM) as well as with ultrahigh-resolution EELS measurements were performed to study QD-decorated NWs structures and to measure the bandgap differences between bare GaN NWs and QD-decorated NWs. STEM imaging or spectrum imaging (SI) measurements were performed at 80 kV using Titan Themis Z TEM (Thermo Fisher, formerly known as FEI Co, USA) operating in the 40–300 kV range, which was equipped with a double Cs (spherical aberration) corrector, a high brightness electron gun (x-FEG), an electron beam monochromator, and a Gatan Quantum 966 imaging filter (GIF). The low-loss spectra were acquired in the so-called microprobe STEM mode with about 1 mrad semi-convergence angle (4 nm probe size). The electron beam monochromator operation was optimized by adopting the method first implemented by Govyadinov and colleagues[243] and described in detail by Lopatin et al.[244]

8.2.4. Optical Characterization

UV-vis Varian Cary 5000 spectrophotometer was employed to evaluate the p-MnO QDs and GaN NW bandgaps through RT absorption measurements. Material optical properties before and after QD drop-casting on NWs were examined by RT micro-photoluminescence (μ-PL) measurements using Horiba LabRAM Aramis Raman spectrometer attached to Kimmon Koha continuous-wave (CW) He-Cd laser (λ = 325 nm) and 40× objective. PL measurements of samples before and after QD drop-casting on NWs at RT and 5 K were performed at 5 K and RT by a 325 nm CW He–Cd laser and a 244 nm (the double frequency of 488 nm line) Ar^+ LEXEL laser, whereby PL signals were captured by an Andor monochromator connected to a charge-coupled device camera. In addition, to obtain RT PL excitation (PLE) spectra, Edinburgh Instruments FLS980 spectrometer attached to a 1000 W Xe lamp (Newport) was used. Note that no PL measurements were acquired within the first 30 minutes after the drop-casting process MnO QDs in ethanol on GaN NWs to ensure that ethanol has evaporated completely based on our examinations. Time-resolved photoluminescence (TRPL) was performed to understand the ultrafast carrier dynamics of the carriers before and after QD drop-casting on NWs at RT and 5 K. A third harmonic line of 266 nm (obtained by APE-SHG/THG) of the Coherent Mira 900 Ti: sapphire femtosecond laser (using a fundamental line of 800 nm) was used for excitation. To capture the long TRPL decay (>2 ns) of the sample, the laser repetition rate was reduced to 2 MHz by using a pulse selection (APE). To detect the TRPL signal, the streak camera (Hamamatsu C6860) was employed in single-sweep mode.

8.3.   Results and Discussion

Figure 8.1a and b show cross-sectional and top SEM view of one of the GaN NW samples grown on Si substrate, revealing their average length of ~750 nm and ~35 nm average diameter, while the other GaN NW sample is characterized by a ~3.5 μm and ~50 nm average length and diameter, respectively. The HR-TEM image shown in Figure 8.2a indicates that the solution-processed p-MnO QDs are highly crystalline and their diameter ranges from 3.5 to 7 nm. The STEM image of a single GaN NW decorated by MnO QDs through the drop-casting method is shown in Figure 8.2b (note that Figure 8.2b suggests that some QDs attached to the NWs are clustered). According to our previously reported XPS findings, the MnO phase is dominant (81.5%) in the composition of these p-MnO QDs, with much smaller contributions from MnOOH (12.0%) and Mn_2O_3 (6.5%), resulting in the unique characteristics discussed in our previous work.[95, 245] In the same study, we demonstrated their p-type conductivity using a field-effect transistor (FET) and Kelvin probe measurements.[95]

To determine the bandgap of the p-MnO QDs, PL, PLE, and absorption measurements were performed at RT. Figure 8.3a shows the corrected PL emission from the p-MnO QDs. The corrected PLE and absorption spectra of MnO QDs revealed a wider bandgap range (~5 eV) than that of GaN (3.5 eV), as shown in Figure 8.3b and c respectively, and agrees with the previous literature.[95] A slight tail shown in the 4.65–4.3 eV (265–285 nm) range can be related to surface states in QDs, as discussed in our recent reports.[95, 17] It should be noted that both the absorption and the PLE spectra have been corrected by subtracting the ethanol signal (The PL and PLE spectra of ethanol are provided in Figure 8.4a and b).

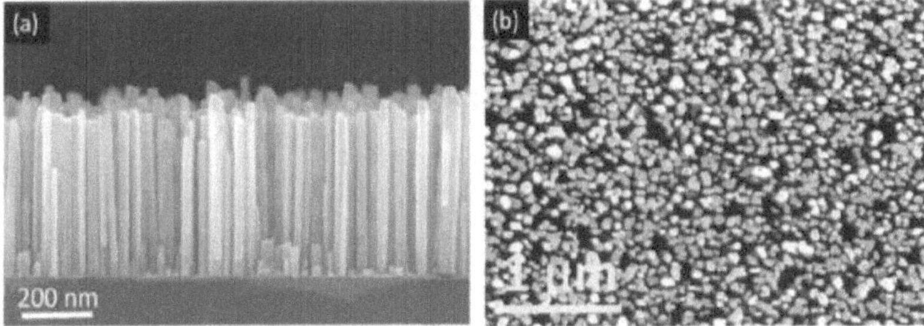

Figure 8.1. (a) Cross-sectional SEM image of GaN NWs and (b) the top view of the same SEM image.

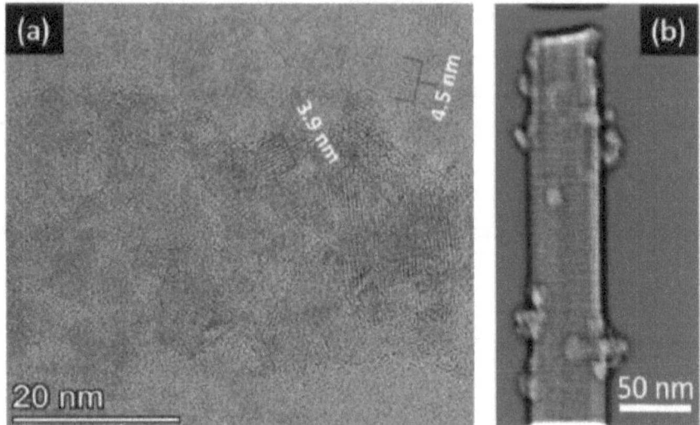

Figure 8.2. (a) HR-TEM images of MnO QDs in a TEM grid and (b) STEM image of MnO QD-decorated GaN NWs. (The image is obtained using a monochromated electron beam with a very small convergence angle to improve the EELS energy resolution).

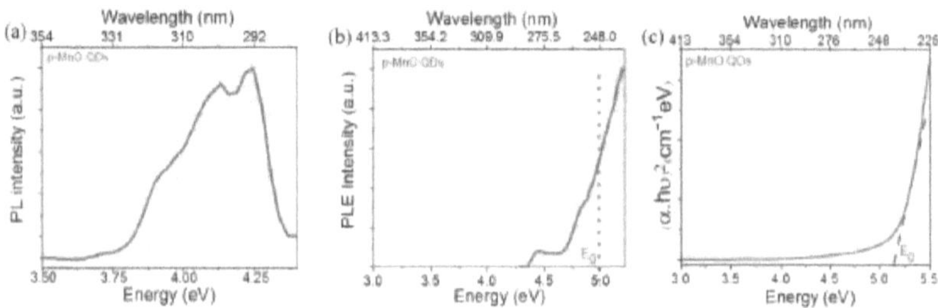

Figure 8.3. (a) The corrected PL spectrum of p-MnO QDs after subtracting the ethanol signal. (b) The corrected PLE spectra of MnO QDs in ethanol after subtracting the ethanol signal. (c) Tauc absorption plot of MnO QDs in ethanol at RT.

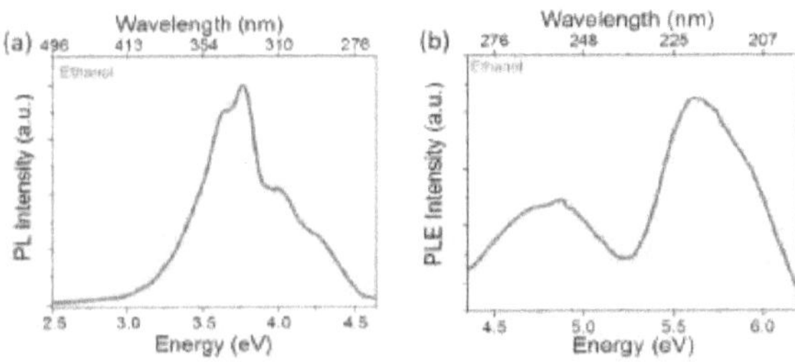

Figure 8.4. (a) RT PL spectrum of ethanol excited by 230 nm and (b) RT PLE spectrum of ethanol at 320 nm.

To study the effect of MnO QD functionalization on GaN emission, we investigate the PL and μ-PL emission produced by GaN before and after functionalizing NWs with QDs. Figure 8.5a shows the GaN near band edge emission (NBE) at 363.5 nm (3.41 eV) before and after MnO QD drop-casting at RT excited by 244 nm (corresponding to the energy above the MnO bandgap) as well as μ-PL excited by 325 nm as depicts in Figure 8.5b. Notably, no MnO emission from QD-decorated NWs is observed even when the samples were excited by $\lambda = 244$ nm. The PL (and μ-PL by 325 nm) spectra shown in Figure 8.5a and b demonstrate up to 2.8-fold integrated intensity enhancement at RT is achieved in the QD-decorated GaN NW emission compared to that produced by bare GaN NWs (before QD drop-casting), whereas no peak shift is observed. Worth mentioning, a very weak yellow band is observed, indicating the superior quality of the GaN NWs. To confirm this emission enhancement, we repeated RT μ-PL measurements of QD-decorated NWs by recording the PL spectrum at different time intervals after the drop-casting process, for up to three hours. These findings are reported in Figure 8.6a and b, which demonstrates that there is no significant change in the emission intensity of QD-decorated NWs over time, confirming the consistency of our results.

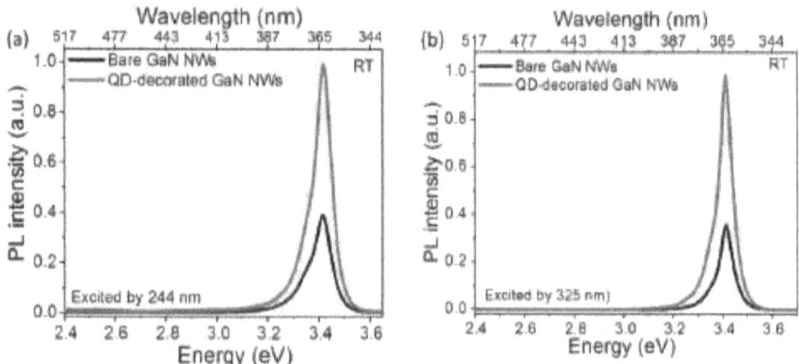

Figure 8.5. (a) RT PL spectra of bare GaN NWs and QD-decorated NWs excited by 244 nm. (b) RT μ-PL spectra of bare GaN NWs and QD-decorated NWs when excited by a laser at λ = 325 nm.

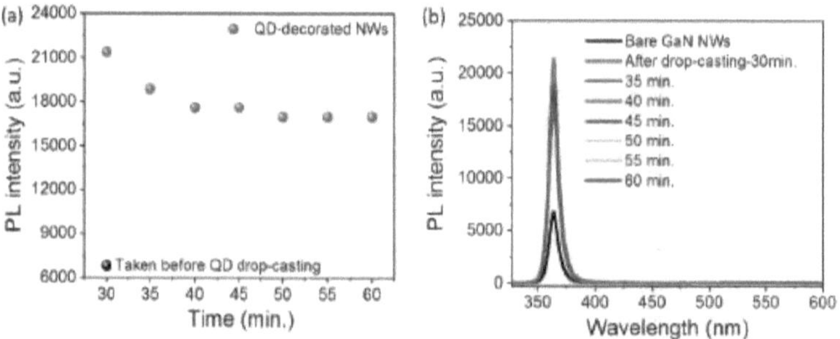

Figure 8.6. (a) The PL intensity of QD-decorated GaN NWs as a function of the time after the drop-casting process, PL signal was recorded at several time intervals after drop-casting QDs on GaN NWs compared to bare NWs. (b) The corresponding PL spectra of bare GaN NW.

Figure 8.7a shows typical PL spectra produced by bare and QD-decorated NWs at low temperature (at 5 K), compared to RT. At 5 K, there are two dominant peaks. The most intense peak (Peak 1) is located at ~ 3.47 eV (357 nm), which was attributed to the common GaN NBE, including free and bound exciton related emissions.[42, 179] This peak is observed at both 5 K and RT (at 360 nm). The second peak (Peak 2) is located at ~3.42 eV (362 nm) at 5 K is also a common GaN peak at low temperature and is dissociated at the RT, which was attributed to exciton bound to stacking fault.[42, 179] Figure 8.7a shows that, the PL emission of QD-decorated NWs considerably enhanced at both RT and 5 K, which is in line with the RT μ-PL measurements. Figure 8.7b shows higher resolution PL spectra, confirming that the PL integrated intensities of Peak1 and Peak 2 increases by ~3.66-fold and ~3.9-fold, respectively, after MnO QD drop-casting compared to bare GaN NWs. We attribute the greater enhancement at 5 K to the fact that the electron-hole recombination process is dominated by radiative recombination at low temperatures. These findings show that PL measurements indicate that, after QD functionalization, the NW emission increases by one order of magnitude compared to bare NWs, which is superior to the < 2-fold enhancement reported for NWs surface-passivated by KOH solution.[233]

To rule out any laser or ethanol effects on the PL enhancement, we studied extensively their effect on bare GaN NWs as a function of time. As shown in Figure 8.8a, the PL intensity of bare GaN remained unchanged after prolonged laser exposure, recorded at several time intervals. Similarly, after drop-casting ethanol only on bare NWs, no significant change is observed in the GaN emission intensity, as shown in Figure 8.8b. Thus, these results confirm that QDs are solely responsible for enhancing the GaN emission as depicted in Figure 8.9.

The emission enhancement of QD-decorated NWs was optimized based on the QD density, as we observed that the GaN emission intensity decreased after several drop-casting cycles. μ-PL measurements (shown in Figure 8.10a) show that the optimized density yields ~ 2.8-fold NBE intensity enhancement compared to that of bare NWs, which declines to ~2-fold after several cycles and remains at this level thereafter. The FLAL synthesis method adopted in this work does not permit estimating the QD density after each cycle. However, the SEM image (Figure 3b) of NWs obtained after several drop-casting cycles reveal that when we performed more than three drop-casting cycles, we observed the formation of a densely packed QD-layer, and a few NWs were toppled over as can be seen in Figure 3b due to drop-casting process. This could result in excessive scattering of emitted light, due to which only a small portion of emitted light can be collected. In addition, most likely, the QD clusters increase carrier scattering. Similar behavior has been observed in LEDs based on QDs; the emission intensity decreases as the QD density increases beyond a certain threshold. For efficient transport, low-density QDs are preferable to well-distributed QDs and can be obtained by both increasing the applied field, and by lowering the barrier between dots. [245-247]

To further elucidate the QD effect on GaN efficiency, we estimated the internal quantum efficiency (IQE) for both bare and QD-decorated NWs. In this case, the IQE represents the ratio of the integrated PL intensity of GaN NWs at 5 K to that at RT. We considered maximum efficiency (100%) at 5 K by assuming that the non-radiative centers are frozen at low temperature (according to the Rashba's treatment).[56, 180, 248] We estimated the IQE for NBE emission (Peak 1), as Peak 2 was absent from the spectrum obtained at RT from PL spectra shown in Figure 8.7a. The IQE value increases from ~28% for bare GaN NWs to ~46% when GaN NWs are hybridized with QDs, which confirms that radiative recombination contribution increases after QD functionalization. These results demonstrated that a considerable improvement in the GaN optical efficiency is achieved by functionalizing NWs with QDs.

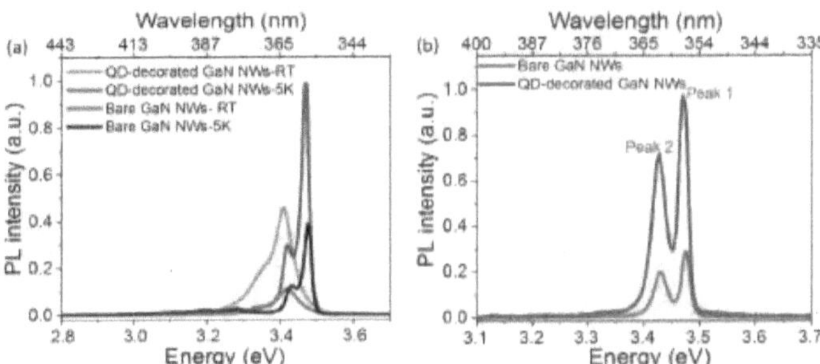

Figure 8.7. (a) Comparison of PL spectra at low temperature and RT for both bare GaN and QD-decorated GaN NWs and (b) Higher resolution PL spectra at 5K for bare GaN and QD-decorated GaN NWs.

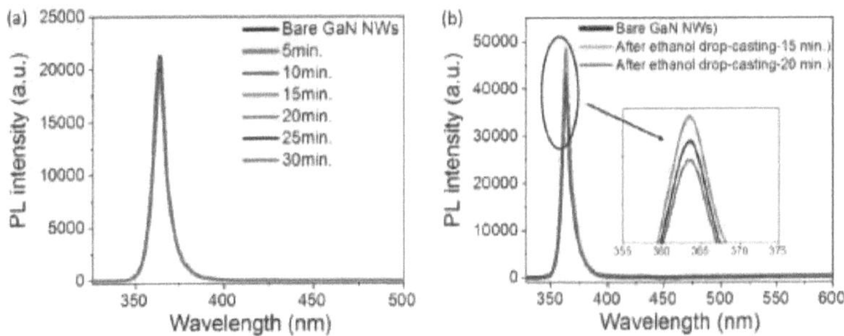

Figure 8.8. (a) PL intensity of bare GaN NWs as a function of time at 5 min intervals, confirming no laser effect on NW emission is observed. (b) PL spectra were taken at different periods; no ethanol effect is observed.

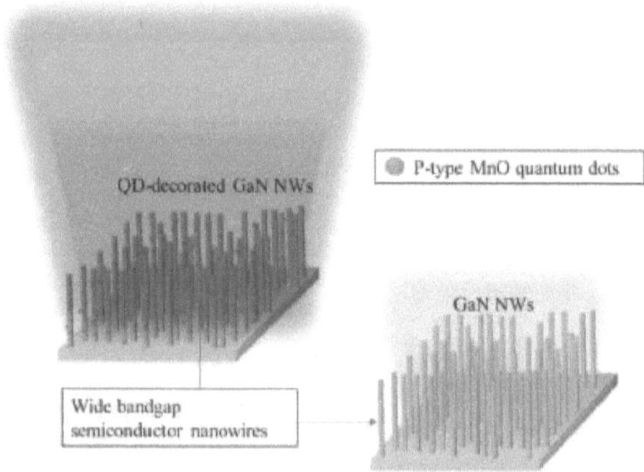

Figure 8.9. Schematic diagram of GaN NWs emissions without and with QDs.

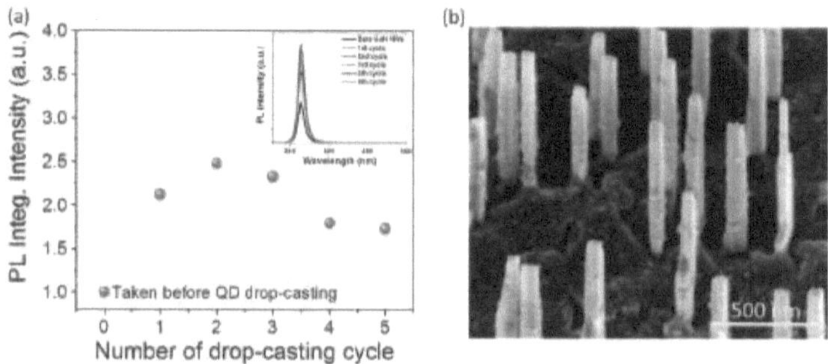

Figure 8.10. (a) PL intensity of GaN NWs as a function of QD density (drop-casting cycles), (the inset shows the corresponding PL spectra based on the drop-casting cycle). (b) SEM image of GaN NWs after three drop-casting cycles of MnO in ethanol.

To investigate the mechanisms underlying the observed emission enhancement, we studied the PLE and ultrahigh-resolution EELS measurements of the QD-decorated GaN NW samples. Figure 8.11a shows the PLE spectra in close proximity to the NBE of GaN NWs (at 370 nm) before and after QD drop-casting. It is evident that the GaN PLE spectrum is enhanced in the deep UV range from 250 nm (4.95 eV) to 340 nm (3.6 eV) after functionalizing the GaN NWs with QDs, while the GaN bandgap remains at 354 nm. The greatest PLE signal enhancement (~2-fold) occurs at > 250 nm (4.95 eV, which is close to the QD bandgap). This value is similar to the enhancement ratio of the PL integrated peak intensity after functionalizing NWs with QDs.

Ultrahigh-resolution EELS measurements at very low loss energy (200 meV) were also conducted and the results are shown in Figure 8.11b. For this purpose, the spatial variation of the bandgap was mapped with the energy resolution of about 45−50 meV, defined as the full width at half maximum (FWHM) of the zero energy loss peak (ZLP).[243, 244] Moreover, the background subtraction was performed by fitting the ZLP measured in a vacuum. The ultrahigh-resolution EELS analysis demonstrates that QDs absorb some energy, as indicated by "glowing" QDs in the EELS map in Figure 8.11c. The EELS signal in the energy range above the bandgap produced by bare GaN differs from that obtained for QD-decorated GaN NWs. Specifically, the signals shown in Figure 8.11b indicate that, for both bare NWs and QD-decorated NWs, the onset of the conduction band transition (bandgap width) starts at about the same energy level (~3.4 eV), indicating the NW bandgap,[243] which is in line with PLE results. However, at 4.95 eV (corresponding to λ = 250 nm, which is close to the QD bandgap) the signal grows faster for QD-decorated NWs compared to the bare ones. The EELS findings indicate a significant increase in the density of states of the QD-decorated GaN NWs due to QD functionalization, compared to bare GaN NWs,[243, 244] thus confirming the PL and PLE results. Worth noting, no plasmonic effect has been observed by EELS as it is expected because the QDs are wide bandgap semiconductor material.

The energy band diagram of the GaN and p-MnO QDs sample is shown in Figure 8.12a, confirming a good bandgap overlap between GaN and p-MnO QDs. Moreover, based on the PLE and EELS findings, the PL enhancement of QD-decorated NWs can be attributed to the energy transfer from MnO QDs to GaN due to the bandgap difference. In this case, for QD-decorated NWs, after exciting QDs and creating electron–hole (e–h) pairs, the energy resulted from e–h recombination process is transferred to the GaN NW band to excite more electrons from the valence band to the conduction band, creating higher e-h pair density (i.e. ET process occurred from higher bandgap energy to lower bandgap energy, as shown in Figure 8.12b.). In this case, the GaN carriers are excited through two paths: directly by laser and via ET process, resulting in more radiative-recombination processes[249] than that in bare GaN NWs as the band alignment is shown in Figure 8.12a facilitates such ET process. Specially, there is energy resonance between the bandedge tail (shown below the MnO bandedge) and the GaN bandedge, as indicated by the PLE and absorption measurements of MnO QDs. This ET phenomenon has been observed previously, but the ET direction was reversed (i.e., from III-nitride to smaller-bandgap QDs),[60, 250] facilitating the development of visible LED. However, ET from wider-bandgap QDs to GaN has never been reported.

To elucidate our ET transfer hypothesis, we carried out time-resolved PL (TRPL) measurements before and after functionalizing NWs by QDs. We analyze the carrier lifetime (τ_{PL}) for Peak 2 of GaN NWs at low temperature (5K) as this peak emerges at low temperature, implying that the radiative-recombination process is dominant for this peak.

In addition, at low temperature, it was assumed that most non-radiative recombination centers are frozen.[169, 180, 248] τ_{PL} of Peak 2 decreases from 2.03 ns (for bare GaN NWs) to 1.39 ns (for QD-decorated GaN), as shown in Figure 8.13a, resulting in a ~0.68 ratio of τ_{PL} after and before decoration with QDs, which implies that the radiative recombination rate increases after QDs functionalization.[181, 251]

To further confirm our ET hypothesis, we functionalized ZnO NTs grown on Si by PLD (~ 3.7 eV bandgap) by the MnO QDs using the same drop-casting process. The growth and properties of these NTs have been published elsewhere;[175] and we found that the emission intensity increases by more than 4-fold as shown in Figure 8.13b. This is evidence supporting our assertion that depositing wider-bandgap QDs on the surface of smaller-bandgap semiconductor nanostructures can lead to semiconductor enhancement. As this is a novel study, additional investigations may be needed to elucidate the underlying mechanisms. This novel finding has important implications for future efforts aimed at utilizing wide-bandgap QD materials to obtain devices operating in the deep UV range. We expect that, in the QD-decorated NW-based LED structure, the turn-on voltage is reduced due to greater carrier density resulting in higher EL intensity compared to that from bare NWs. Also, an anti-reflective coating may be needed to further improve the light extracting efficiency, leading to high-efficiency UV light-emitting devices.[242, 252]

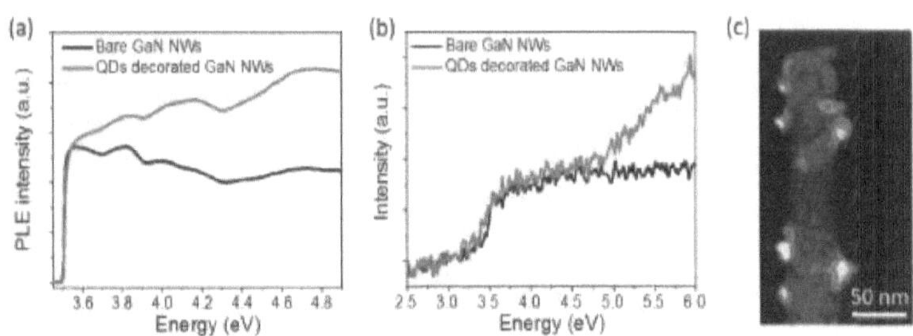

Figure 8.11. (a) PLE spectra of bare GaN NWs and those decorated with QDs. (b) Comparison of ultrahigh-resolution low-loss EELS signal (after background subtraction) obtained from bare GaN NWs with that produced by GaN NWs decorated with QDs. (c) Fitting map of the low-loss EELS signal obtained from GaN NWs vs. signal produced by QDs.

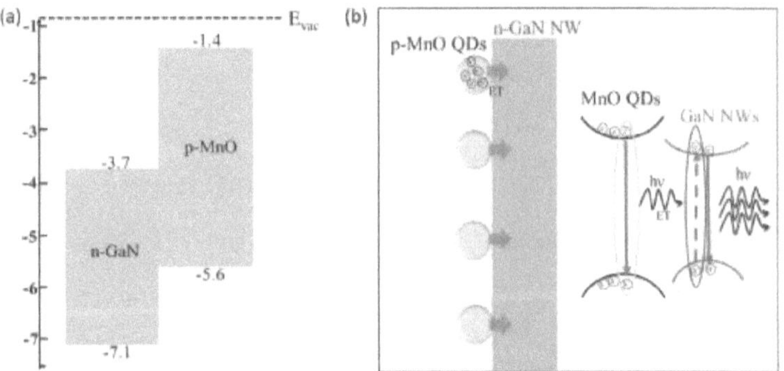

Figure 8.12. (a) Band diagram of GaN[223] and MnO QDs[95] materials. (b) Schematic diagram of ET from MnO QDs to GaN NWs.

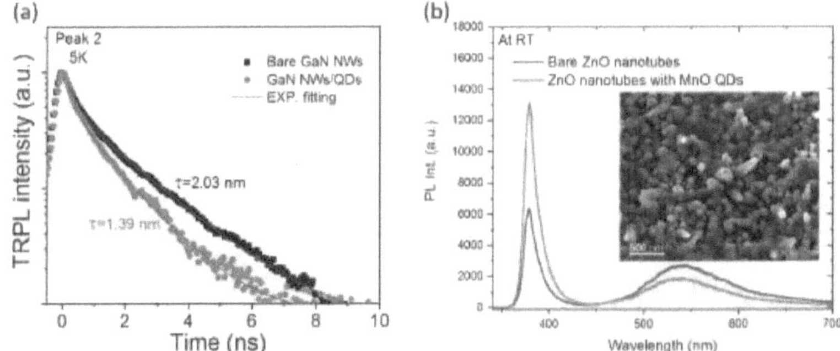

Figure 8.13. (a) Low-temperature TRPL measurements of Peak 2 before and after QD drop-casting, on GaN NWs. The best-fitting reveals a single decay for both curves. (b) RT μ-PL spectra of bare ZnO NTs compared to ZnO NTs functionalized by MnO QDs.

8.4. Summary

In this chapter, I reported on the first study as a part of which the optical UV efficiency of NWs was enhanced significantly by functionalizing them by p-MnO QDs characterized by a wider bandgap to transfer the energy from the QDs to GaN. Our findings show that, when highly crystalline p-MnO QDs are drop-casted on GaN NWs, the NBE emission of GaN is highly (~ 290%) enhanced, i.e., it is ~3.8-fold greater than that produced by bare GaN NWs. A considerable IQE increase obtained after decorating GaN with QDs further demonstrates that the radiative recombination rate increases after decorating NWs with QDs. As this process has never been used or studied for enhancing III-nitride emission, we have succeeded in demonstrating that functionalizing QDs (with 5 eV bandgap) with III-nitrides (including GaN, InGaN, or AlGaN NWs) has the potential to enhance the LED efficiency. EELS and PLE measurements showed enhancement in the deep UV range after functionalizing the GaN NWs, while also revealing that, above 4.9 eV (equivalent to the

QD bandgap), the signal grows faster for QD-decorated NWs above, indicating an increase in the density of states after functionalizing GaN with QDs. These analyses show that energy transfer occurs from MnO to GaN NWs, which is the most likely reason behind the GaN emission enhancement. Thus, this novel environmentally-friendly (with zero waste) strategy would be highly beneficial for state-of-the-art technology development aimed at enhancing the performance of light-emitting devices, as well as in wide-bandgap semiconductor applications, such as transistors, photovoltaic cells, and photodetectors.

Chapter 9

Conclusion and Future Direction

9.1. book Conclusion

Several methods and devices are shown in this book have been demonstrated for the first time. III-nitride (mainly GaN) growth is presently hindered by the dislocation defects due to the lattice mismatch between the material and the substrate as well as low device efficiency as described in Section 2.7. Therefore, as a part of the work reported in this book, these challenges have been addressed, as the growth method based on PLD (which is a low-cost technique) can be applied regardless of the lattice mismatch between GaN and substrate, resulting in high-quality GaN. It was further demonstrated that the PLD can be adopted to produce four-inch wafers, thus meeting the requirements of most commercial applications, including large-scale production. As a part of the present investigation, GaN NWs were successfully grown by PLD, which is a direct method that does not require a catalyst, seeding, or complex fabrication processing, for the first time. The growth mechanism relies on the surface energy and energy of species landed on the substrate, which are the key factors responsible for producing dislocation-free catalyst-free state-of-the-art GaN NWs grown on emerging and the common substrate such as Ga_2O_3, Si, Al_2O_3 (c-sapphire), GaN, graphene, MXene, and TMD. Moreover, STEM findings indicate a complete absence of dislocations between the substrate and the WL due to the polycrystalline nature of WL, while NWs exhibit a single wurtzite crystal structure. Carrier dynamics and optical characterizations on all samples confirmed superior UV efficiency, while a slight droop at high excitation densities is ascribed to a minor contribution from Auger recombination at high carrier injection rates, demonstrating that these NWs can be adopted for obtaining high-efficiency UV LEDs, including vertical emitting devices such as VCSELs. This PL growth strategy can be used as a template for the production of a wide variety of flexible and large-scale applications, as it is unaffected by the lattice mismatch issue. These findings open up new horizons in research on GaN-based emitting devices, including UV vertical LEDs and flexible devices.

Besides, further improvements to the GaN NW performance have been achieved by functionalizing the GaN NWs with different emerging materials, mainly perovskites ($CH_3NH_3PbI_3$ and $CsPbBr_3$) and wide-bandgap semiconductor QDs, for the first time. The perovskite functionalization strategy resulted in significant sensing characteristics of GaN-photodetector. On the other hand, the QD functionalization improves UV efficiency.

The two types of perovskite were hybridized with GaN NWs to fabricate high-performance broadband self-powered photodetector. It was further demonstrated that the photodetector performance was mainly governed by the work function alignment between the perovskite and GaN NWs. Based on the type of contact layers, four photodetectors were presented, and their performance was compared. The vertical photodetector configuration was superior to lateral configuration, while only $CsPbBr_3$/GaN NWs were capable of operating in the self-powered mode. These results are nonetheless encouraging, given that this is the first attempt to develop a photodetector based on GaN NWs/ organic or inorganic perovskite architecture.

The GaN NW functionality was also enhanced by hybridizing it with QDs, thus improving the efficiency of the UV emission significantly. As a part of this process, a proof-of-concept study was conducted, demonstrating that the optical UV efficiency enhancement was due to QDs having a wider bandgap than that of GaN, which allowed energy transfer from the QDs to GaN when they are pumped optically. These findings show that, when wider-bandgap highly crystalline p-MnO QDs are drop-casted on GaN NWs, the PL emission of GaN is highly enhanced, i.e., it is ~3.9-fold greater than that of bare GaN NWs. Further analyses showed that energy transfer occurs from MnO to GaN NWs, which is the most likely reason behind the GaN emission enhancement. This method yields much better emission enhancement than can be attained by NWs passivated by KOH, for example, which is a highly toxic solution. Thus, this novel environmentally-friendly strategy would be highly beneficial for state-of-the-art technology development, especially for light-emitting devices, as well as wide-bandgap semiconductor applications, such as transistors, photovoltaic cells, and photodetectors.

All these strategies were successfully employed to obtain devices with unique optical and structural properties, paving the way for a wide range of novel and flexible applications.

9.2. Future Direction

9.2.1. Material Growth

Work in the field of PLD will continue, as there is still a need to better control and enhance the GaN NW growth process. It is envisaged that, as a part of this endeavor, the growth of GaN NWs with hexagonal flat end under different growth conditions will be investigated to obtain the most optimal conditions. Findings yielded by such studies would be beneficial for improving the efficiency of GaN and III-nitride-based emitting devices, including LDs, as these NWs can be used as a template to fabricate dislocation-free high-efficiency devices.

Furthermore, extensive research on the use of PLD with other III–N materials (such as InN and AlN) and their alloys with GaN (e.g., InGaN and AlGaN) by optimizing the growth parameters for differing applications is also expected. Similarly, the fabrication of LED devices based on the GaN NWs grown on the p-GaN substrate will also be investigated. In the research on InGaN/GaN MQWs, a new method based on a cost-effective technique such as PLD is urgently needed.

9.2.2. Device Fabrication

Light-emitting devices based on QD-functionalized III-nitrides will be an important topic of future studies, as there is a need to demonstrate the effectiveness of the growth method presented herein improving the efficiency of UV LED devices. Fabrication of VCSELs and vertical LED devices based on the III-nitrides NWs discussed here, as well as flexible devices, is one of the most important future work directions, as the NWs developed as a part of this work can be grown on any substrate. In addition, photodetectors based on PLD III-nitride NWs grown on different substrates will also be studied in the future. It would also be beneficial to conduct studies on band energy alignment using new materials for contact layers with a different configuration.

APPENDIX

Publications and patents

1. Iman S Roqan and **Dhaifallah R. Almalawi** "Dislocation Free Semiconductor Nanostructures Grown by Pulsed Laser Deposition with no Seeding or Catalyst" **US patent with number** 62/803,515, February 10, 2019, PPB Ref. 0338-427

2. **Dhaifallah Almalawi,** Sergei Lopatin, Somak Mitra, Tahani Flemban, Alexandra-Madalina Siladie, Bruno Gayral, Bruno Daudin, and Iman S Roqan, Enhanced UV Emission of GaN Nanowires Functionalized by Wider Bandgap Solution-Processed p-MnO Quantum Dots, ACS Appl. Mater. Interfaces, 12, 34058−34064, 2020.

3. **Dhaifallah R. Almalawi**, Sergei Lopatin, Bin Xin, Paul Edwards, Ram C. Subedit, Yi Wan, Nini Wei, Boon S. Ooi, Vincent Tung, Robert W Martin, Iman S. Roqan, Direct growth of high-quality GaN NWs on a wide range of substrates via polycrystalline GaN wetting layer formation, (Submitted to small).

4. **Dhaifallah Almalawi,** Norah Alwadai, Bin Xin, Somak Mitra, Yusin Pak, and Iman S. Roqan, Enhanced Photodetectors based on GaN Nanowires Functionalized by Perovskite, (submitted to Applied physics letters).

5. **Dhaifallah Almalawi,** Jungwook Min, Bin Xin, Boon S. Ooi and Iman Roqan, InGaN/GaN MQW-based LED structure grown on PLD GaN NWs, (to be submitted).

6. Ajia, Idris; **Almalawi, Dhaifallah**; Lu, Yi; Lopatin, Sergei; Li, Xiaohang; Liu, Zhiqiang; Roqan, Iman; Sub-quantum-well influence on carrier dynamics in high-efficiency DUV dislocation-free AlGaN/AlGaN-based multiple-quantum-wells, *ACS Photonics* 7 (7), 1667-1675, 2020.

7. I. A. Ajia, Y. Yamashita, K. Lorenz, M. M. Muhammed, L. Spasevski, **D. Almalawi**, J. Xu, K. Iizuka, Y. Morishima, D. H. Anjum, N. Wei, R. W. Martin, A. Kuramata, and I. S. Roqan,

GaN/AlGaN multiple quantum wells grown on transparent and conductive (-201)- oriented β-Ga2O3 substrate for UV vertical light-emitting devices, Appl. Phys. Lett. 113, 082102,2018.

8. Bin Xin, Yusin Pak, Somak Mitra, **Dhaifallah Almalawi**, Norah Alwadai, Yuhai Zhang, and Iman S. Roqan, Self-Patterned CsPbBr3 Nanocrystals for High-Performance Optoelectronics, ACS Appl. Mater. Interfaces, 11, 5223−5231, 2018.

9. Somak Mitra, Yusin Pak, Naresh Alaal, Mohamed N. Hedhili, **Dhaifallah R. Almalawi**, Norah Alwadai, Kalaivanan Loganathan, Yogeenath Kumarasan, Namsoo Lim, Gun Y. Jung, and Iman S. Roqan, Novel P-Type Wide Bandgap Manganese Oxide Quantum Dots Operating at Deep UV Range for Optoelectronic Devices, Adv. Optical Mater. 7, 1900801, 2019.

10. Somak Mitra, Yusin Pak, Bin Xin, **Dhaifallah R. Almalawi**, Nimer Wehbe, and Iman S. Roqan, Solar-Blind Self-Powered Photodetector Using Solution-Processed Amorphous Core−Shell Gallium Oxide Nanoparticles, ACS Appl. Mater. Interfaces, 11, 38921−38928, 2019.

11. Y. Pak, S. Mitra, N. Alaal, B. Xin, S. Lopatin, **D. Almalawi**, J.-W. Min, H. Kim, W. Kim, G.-Y. Jung, and I. S. Roqan, Dark-current reduction accompanied photocurrent enhancement in p-type MnO quantum-dot decorated n-type 2D-MoS2-based photodetector, Appl. Phys. Lett. 116, 112102, 2020.

12. Somak Mitra, Mufasila Mumthaz Muhammed, Norah Alwadai, **Dhaifallah R. Almalawi**, Bin Xin, Yusin Pak, and Iman S. Roqan, Optimized performance III-nitride-perovskite-based heterojunction photodetector via asymmetric electrode configuration, RSC Adv., 10, 6092-6097, 2020.

13. Bin Xin, Naresh Alaal, Somak Mitra, Ahmad Subahi, Yusin Pak, **Dhaifallah Almalawi**, Norah Alwadai, Sergei Lopatin, Iman S. Roqan, identifying carrier behavior in ultra-thin indirect-bandgap CsPbX3 nanocrystal films for use in UV/visible-blind high-energy detectors, Small, 2004513, 2020.

www.ingramcontent.com/pod-product-compliance
Lightning Source LLC
LaVergne TN
LVHW042118190726
843493LV00006B/1520